Gravity-Driven Water Flow in Networks Theory and Design

Gravity-Driven Water Flow in Networks Theory and Design

Editor

Azhaire Ivanov

Gravity-Driven Water Flow in Networks Theory and Design
Edited by **Azhaire Ivanov**

Printed in 2017

ISBN: 978-1-68117-142-5
Library of Congress Control Number: 2015952747

© 2016 by
SCITUS Academics LLC,
616, Corporate Way, Suite 2, 4766,
Valley Cottage, NY 10989

www.scitusacademics.com

This book contains information obtained from highly regarded resources. Copyright for individual articles remains with the authors as indicated. All chapters are distributed under the terms of the Creative Commons Attribution License, which permits unrestricted use, distribution, and reproduction in any medium, provided the original author and source are credited.

Notice

Reasonable efforts have been made to publish reliable data and views articulated in the chapters are those of the individual contributors, and not necessarily those of the editors or publishers. Editors or publishers are not responsible for the accuracy of the information in the published chapters or consequences of their use. The publisher believes no responsibility for any damage or grievance to the persons or property arising out of the use of any materials, instructions, methods or thoughts in the book. The editors and the publisher have attempted to trace the copyright holders of all material reproduced in this publication and apologize to copyright holders if permission has not been obtained. If any copyright holder has not been acknowledged, please write to us so we may rectify.

Preface

Gravity-driven water flow networks are a crucial method of delivering clean water to millions of people worldwide, and an essential agricultural tool.

This book attempts to bridge the gap between fundamental fluid mechanics and applied and useful technology-based material in the various existing references on gravity-driven water networks. In particular, the topics have been chosen to add simplicity, a sound technical foundation, and support too many of those confined. A benefit of the fundamentals approach is the production of original design graphs, formulas, and computational algorithms for the correct, sustainable designs of single-and multiple-pipe gravity-driven water networks. Both theory and design are covered in this book, along with the analysis that must join the two. It is a considerable challenge to span this range effectively. This book delivers an all-encompassing guide to designing these water networks, combining theory and case studies. It comprises design formulas for water flow in single or multiple, uniform or non-uniform diameter pipe networks; case studies on how systems are built, used, and maintained; comprehensive coverage of pipe materials, pressure ratings, and dimensions. It is a key resource both for working engineers and engineering students and instructors.

Table of Contents

CHAPTER 1 Assessing the Impact of Transitions from Centralised to Decentralised Water Solutions on Existing Infrastructures – Integrated City-Scale Analysis with Vibe 1

HIGHLIGHTS 1
ABSTRACT..1
INTRODUCTION...2
METHODS 4
    Modelling concepts..4
    Generation procedure 5
INTEGRATED SCENARIO ANALYSIS OF WATER INFRASTRUCTURE
....10
    Coupling urban water infrastructure 10
    Performance assessment 11
DEFINITIONS OF SCENARIOS AND CASE STUDIES 13
    Case study water infrastructure Innsbruck..14
RESULTS 15
    Integrated generation process – VIBe city 15
    Impact of transitions from centralised to decentralised water infrastructures 16
    Stability of the water networks 18
MODEL BUILDING EXAMPLE WITH NUMEROUS VIRTUAL CASE STUDIES..20
CONCLUSIONS 22
ACKNOWLEDGEMENTS 23

REFERENCES 23
CITATION 27

CHAPTER 2   Detection of Water Pipes and Leakages in Rural Water
             Supply Networks Using Remote Sensing Techniques..29

INTRODUCTION 29
STUDY AREAS 31
  "Southern Conveyor Project" 32
  "Lakatameia" Pipeline 33
  "Frenaros – Choirokoitia" Water Pipe 34
METHODOLOGY 37
RESOURCES 38
  High Resolution Satellite Data 38
  Spectroradiometric Data 38
  Medium Resolution Satellite Data 42
RESULTS 43
  "Southern Conveyor Project" Pipeline 43
  "Lakatameia" Pipeline 46
  "Frenaros – Choirokoitia " Water Pipe 49
DISCUSSION AND REMARKS 54
ACKNOWLEDGEMENTS 54
REFERENCES 55
CITATION 58

CHAPTER 3   An Analysis of the Interface between Evolutionary
             Algorithm Operators and Problem Features for Water
             Resources Problems. A Case Study in Water
             Distribution Network Design ☆ 59

HIGHLIGHTS 59
ABSTRACT 59
INTRODUCTION 60
  Water distribution network design problem 61
  Optimisation of WDNs 62
  Performance analysis 62
  Problem and operator linkage 63

METHOD	64
A common approach to comparing methods	64
Method for characterising optimisers	65
Defining problem features	67
Artificial problems	67
Characterising & selection of optimisers	68
THE WATER DISTRIBUTION NETWORK DESIGN PROBLEM	68
WDN features	70
EVOLUTIONARY ALGORITHM	72
Genetic operators	73
EXPERIMENTAL SETUP	74
Benchmark problems	76
Parameter structure	76
Experimental settings	76
RESULTS	77
General performance	77
Convergence rates	78
Loops and branches	79
Multiple sources, pumps and valves	80
Combinations of operators	81
Benchmark problems	85
CONCLUSIONS	87
ACKNOWLEDGEMENTS	88
REFERENCES	88
CITATION	91

CHAPTER 4   Vadose Zone Heterogeneity Effect on Unsaturated Water Flow Modeling at Meso-Scale 93

ABSTRACT	93
INTRODUCTION	94
MATERIALS AND METHODS	95
Site location	95
GPR Study	96
Numerical modeling of flow	98
RESULTS AND DISCUSSION	100
Architectural characterization of the Studied Sections	100
Modeling water Redistribution during Drainage	104
Modeling water Infiltration during Rainfall	106

Modeling water Infiltration for Meteorological data 109
Link between flow and Geometrical indicators of heterogeneous
Sections 112
CONCLUSION 113
ACKNOWLEDGEMENTS 114
REFERENCES 114
CITATION 116

CHAPTER 5 Hydrodynamic Performances of Air-Water Flows in Gullies with and Without Swirl Generation Vanes for Drainage Systems of Buildings 117

ABSTRACT 117
INTRODUCTION 118
RESEARCH METHODS 120
   Experimental Apparatus and Test Details 120
   Numerical Method and Simulation Details 123
RESULTS AND DISCUSSION 124
   Flow Structures 124
   Air Entrainments by Entry Vortex 128
   Air Bubble Drifts in Test Gullies with/without SGV 132
   Self-Depuration Performances 133
CONCLUSIONS 136
AUTHOR CONTRIBUTIONS 137
REFERENCES 137
CITATION 138

CHAPTER 6 Flexible Heat Pipes with Integrated Bioinspired Design 139

ABSTRACT 139
INTRODUCTION 139
EXPERIMENTAL 141
   Materials 141
   Preparation of wick 142
   Fabrication of heat pipes 142
   Characterization and property measurement 142
RESULTS AND DISCUSSION 144
CONCLUSIONS 150

ACKNOWLEDGMENT	150
REFERENCES	151
CITATION	151

CHAPTER 7 Rehabilitation Priority Determination of Water Pipes Based on Hydraulic Importance 153

ABSTRACT	153
INTRODUCTION	154
METHODOLOGIES	157
Pipe Deterioration	158
Hydraulic Importance: Single Pipe Failure	160
Hydraulic Importance: Multiple Pipe Failures	162
Weights of Attributes	163
Weighted Utopian Approach	166
APPLICATION RESULTS	169
Results: Deterioration Rate	170
Results: Hydraulic Importance of Single Pipe Failure	172
Results: Hydraulic Importance of Multiple Pipe Failures	173
Results for Final Rehabilitation Priority Order	175
CONCLUSIONS	179
ACKNOWLEDGMENTS	180
AUTHOR CONTRIBUTIONS	180
REFERENCES	180
CITATION	183

CHAPTER 8 Evaluation of Actions for Better Water Supply and Demand Management in Fayoum, Egypt Using Ribasim 185

ABSTRACT	185
INTRODUCTION	186
METHODOLOGY	188
The study area	188
RIBASIM description and model capability	190
Schematization of RIBASIM	191
Schematization of water system of Fayoum Governorate	191
Simulated scenarios	193
RESULTS	199
CONCLUSIONS AND RECOMMENDATIONS	203

REFERENCES 204
CITATION 205

CHAPTER 9   Development and Uptake of Scenarios to Support Water
             Resources Planning, Development and Management –
             Examples from South Africa 207

INTRODUCTION 207
THE ABILITY OF SCENARIOS TO ACHIEVE IMPACT IN AN
UNCERTAIN WORLD WITH A FOCUS ON WATER PLANNING,
DEVELOPMENT AND MANAGEMENT 211
    Scenarios and Their Importance in the Water Sector 211
    Some South African Scenarios: Overview and Impact 213
    Dissemination and Impact 219
    Dinokeng Scenarios 220
    The Impact of Scenarios 226
CONCLUSIONS 231
ACKNOWLEDGEMENTS 234
REFERENCES 234
CITATION 238

Index 239

CHAPTER 1

Assessing the Impact of Transitions from Centralised to Decentralised Water Solutions on Existing Infrastructures – Integrated City-Scale Analysis with Vibe

Robert Sitzenfrei, , , Michael Möderl, Wolfgang Rauch

Unit of Environmental Engineering, Institute of Infrastructure, University of Innsbruck, Technikerstr. 13, 6020 Innsbruck, Austria

HIGHLIGHTS

- Integrated analysis of transition from central to decentralised urban water systems.
- Innovative stochastic generation of city-scale case studies successfully applied.
- Coupling water distribution and urban drainage simulations via population.
- Analysis of 81 test cases to identify stable operation ranges for water systems.
- Development and application of a model to predict shear stress performance, $R^2 = 0.99$.

ABSTRACT

Traditional urban water management relies on central organised infrastructure, the most important being the drainage network and the water distribution network. To meet upcoming challenges such as climate change, the rapid growth and shrinking of cities and water scarcity, water infrastructure needs to be more flexible, adaptable and sustainable (e.g., sustainable urban drainage systems, SUDS; water sensitive urban design, WSUD; low impact development, LID; best management practice, BMP). The common feature of all solutions is the push from a central solution to a decentralised solution in urban water management. This approach opens up a variety of technical and socio-economic issues, but until now, a comprehensive assessment of the impact has not been made. This absence is most likely attributable to the lack of case studies, and the availability of adequate models is usually limited because of the time- and cost-intensive preparation phase. Thus, the results of the analysis are based on a few cases and can hardly be transferred to other boundary conditions. VIBe (Virtual Infrastructure Benchmarking) is a tool for the stochastic generation of urban water systems at the city scale for case study research. With the generated data sets, an integrated city-scale analysis can be performed. With this approach, we are able to draw conclusions regarding the technical effect of the transition from existing central to decentralised urban water systems. In addition, it is shown how virtual data sets can assist with the model building process. A simple model to predict the shear stress performance due to changes in dry weather flow production is developed and tested.

INTRODUCTION

Traditional urban water management relies on central organised infrastructure, the most important being the drainage network and the water distribution network. To meet new challenges, such as climate change and changes in the population and land use (growth as well as shrinkage in the cities), it is commonly agreed that water infrastructure needs to be more flexible, adaptable and sustainable (e.g., Brown et al., 2009 and Domènech and Saurí, 2010). These efforts towards increased sustainability are denoted sustainable urban drainage systems, SUDS; water sensitive urban design, WSUD; low impact development, LID; and best management practice, BMP (e.g., Ole et al., 2012). The common feature of all solutions is the push from a central solution to a decentralised solution in urban water management. This transition opens up a variety of technical and socio-economic issues, but until now, a comprehensive assessment of the impact was missing. This absence is mostly attributable to the lack of case studies, and the availability of adequate models is usually limited because of the time- and cost-intensive preparation phase. This preparation includes, among other things, data collection, digitalisation, model construction and calibration (Refsgaard and Henriksen, 2004). However, findings based on a single or just a few case studies are very case specific, and the conclusions can hardly be generalised. In this paper, we present an integrated city-scale analysis based on virtual generated urban water systems. With this approach, we are able to form conclusions about the technical effect of the transition from central to decentralised urban water systems.

Virtual case studies have been created and used in various studies (e.g., Achleitner et al., 2007, Borsanyi et al., 2008, Butler and Schütze, 2005, De Toffol et al., 2007, Lau et al., 2002, Meirlaen et al.,

2001, Schütze et al., 2002 and Rauch et al., 2003) to allow for detailed investigations, even if data are not available. These studies' properties are intended to resemble real-world characteristics, and the case studies are thus referred to as either synthetic, artificial, (semi-)hypothetical or virtual (Sitzenfrei, 2010).

In the field of water distribution system analysis (WDSA), the three example networks provided with the hydraulic solver EPANET2 (Rossman, 2000) have been used in numerous investigations as benchmark systems. Such benchmark systems serve to test the coherence of numerical solvers, as well as to allow for the comparison of methods and solutions. Other established benchmark systems of WDSA are the New York Tunnels system, the Two-Loop system, the Anytown network and the Hanoi network (Sitzenfrei et al., 2013). In the field of urban drainage modelling (UDM), the use of benchmark systems is less established compared to WDSA. However, several benchmark systems were introduced (e.g., Schilling, 1989 and Schütze et al., 1999) to test new approaches (e.g., Rauch and Harremoës, 1999 and Zacharof et al., 2004).

Because of the lack of data regarding the accurate description of ageing and deterioration of urban drainage systems, a network condition simulator (NetCoS) was developed and used in Scheidegger et al. (2011) and Scheidegger and Maurer (2012), respectively. Blumensaat et al. (2012) developed a model for generating sewer models under minimum data requirements, which are to be seen as semi-synthetic case studies. Furthermore, stochastic approaches were developed for the algorithmic generation of conceptual, simplified network models (water distribution and combined sewer systems) based on the variation of layout and size, among other network characteristics (Möderl et al., 2007, Möderl et al., 2009, Möderl et al., 2011 and Trifunovic et al., 2012). Each of these approaches provides an interface to a hydraulic solver (for WDSA: EPANET2, Rossman, 2000; for UDM: SWMM, Rossman, 2004 and Gironás et al., 2010) to analyse the hydraulic performance of the generated networks. However, all of the approaches mentioned above neglect the urban structures (population densities, land use) in the network design and only add the "city layout" after the network generation.

However, for an integrated assessment of transition effects, considering the dynamics at the city scale is essential. In this paper, a novel concept for the generation of virtual urban water systems is presented. In VIBe (Virtual Infrastructure Benchmarking), the urban structure ("virtual city" including population densities, land use and topography) is generated first. Second, the infrastructure networks (including interfaces to hydraulic simulation software) are created according to the state-of-the-art design rules and meet the requirements of the "virtual city". Therefore, the generated infrastructure networks can be spatially linked via the population.

VIBe is based on a number of sub-modules that have been described earlier, i.e., urban structure generation (Sitzenfrei et al., 2010a), sewer generation (Urich et al., 2010) and the water distribution system generation (Sitzenfrei et al., 2010c). In this work, it is shown, how these approaches can be coupled for the integrated analysis of the urban water system. This approach allows the transition of present systems towards decentralised implementations to be analysed (e.g., biofiltration systems, rainwater harvesting, water reuse and infiltration) (e.g., Le Coustumer et al., 2012, Ward et al., 2012, Moore et al., 2012, Dong et al., 2012, Peter-Varbanets et al., 2009 and Barton and Argue, 2009) with respect to the impact on the hydraulics, water quality and emissions performance of the existing centralised systems. In addition, it is shown how virtual data can assist in model building. Therefore, a simple model to predict shear stress performance due to changes in dry weather flow production is developed and tested.

METHODS

In this section, the modelling concepts of the VIBe approach are described (Section 2.1). Further, the generation process of city-scale case studies (Section 2.2) is discussed. How the urban structures (Section 2.2.1) (Sitzenfrei et al., 2010a), the sewer systems (Section 2.2.2) (Urich et al., 2010) and the water distribution systems (Section 2.2.3) (Sitzenfrei et al., 2010c) are generated with the VIBe approach is outlined.

Modelling concepts
VIBe algorithmically generates numerous case studies of urban water systems at the city scale, including infrastructure networks. Therefore, numerous scenarios with different characteristics can be created for further statistical investigations. To provide a hydraulic performance analysis for the generated infrastructure systems, interfaces to external hydraulic simulation software are implemented in the VIBe approach. This allows for generic findings from the statistical evaluations of the results. This helps to identify system coherences, to determine potentials and to enhance the understanding of generated and real-world systems (see Fig. 1). Furthermore, new software, hypotheses or modelling approaches can be tested, e.g., the potential of real time control, urine

separation, rainwater infiltration, optimisation algorithms, parallel modelling algorithms, transition of water savings scenarios.

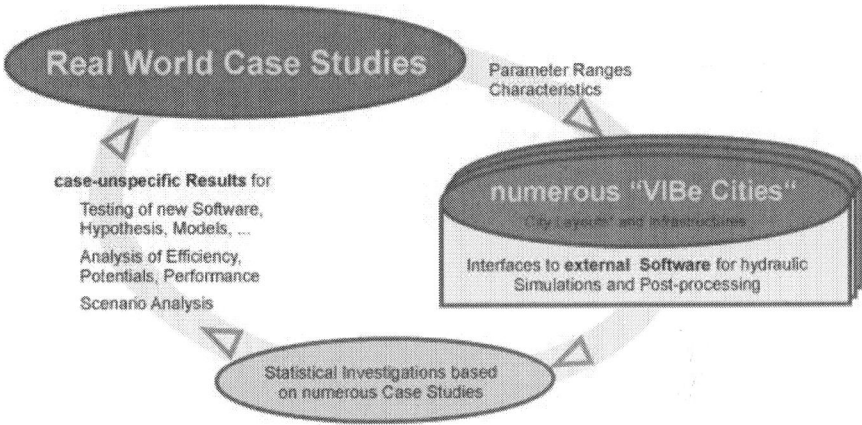

Figure 1. Modelling concept of VIBe.

Generation procedure

The required data are identified from the literature and from real-world case studies, and the parameter ranges are subsequently determined. The "city layout" is created with the urban structure module (US-Module) and by stochastic sampling within the parameter ranges. With predefined topographic boundary conditions (e.g., location in an Alpine valley or size of urban agglomeration), the US-Module generates the digital description of a virtual city (see Section 2.2.1).

The infrastructure modules generate water networks based on the virtual "city layouts" (see Fig. 2). Hence, in the generation process, the "city layout" is created first, represented as GIS data (raster data).

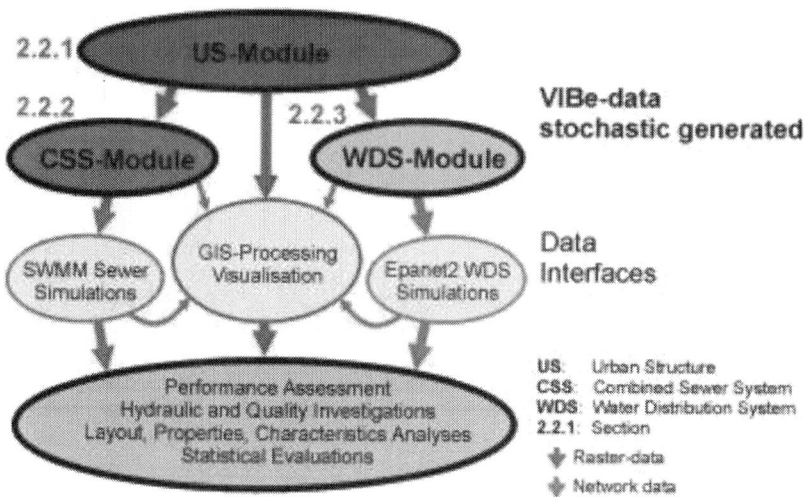

Figure 2. Concept of the generation process in VIBe.

Subsequently, the current implemented infrastructure modules, both the combined sewer system module (CSS-Module, see Section 2.2.2) and the water distribution system module (WDS-Module, see Section 2.2.3) create (with stochastic layout variations) infrastructure systems according to state-of-the-art design rules and stochastic under- or overdesign. For the US-Module, interfaces for GIS-processing and interfaces to external hydraulic modelling tools are supplied. The concept introduced in VIBe has been developed for urban water systems but can be applied to other urban infrastructure (e.g., energy supply, gas network and district heating).

Urban structure ("city") module (US-Module)
The US-Module creates entire cities but without water infrastructure. The generated model builds upon the properties and characteristics of real topography and river systems, and in this study, the model is based on an analysis of an Alpine region. Such a region is characterised by a flat valley floor and a river which meanders (Sitzenfrei et al., 2010a). The US-Module provides all required data for the infrastructure modules with characteristics that are comparable with Alpine real cities but the approach can also be enhanced for other topographies. Furthermore, GIS maps required for water infrastructure systems, such as impervious area, dry weather flow, water demand, and aquifers, are provided by the US-Module (see also Fig. 3) and can be exported via an interface to GIS-processing software.

METHODS

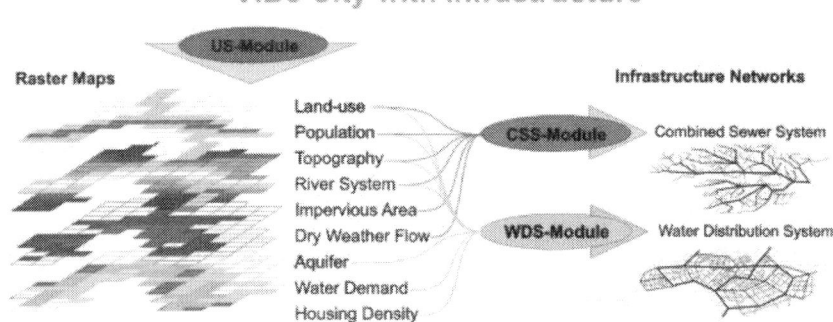

Figure 3. Creation of a VIBe city.

Combined sewer system module (CSS-Module)
The CSS-Module in VIBe generates combined sewer systems based on GIS data created either with the US-Module (see Figs. 3 and 4) or real-world data. This module generates layouts of sewer systems meeting the requirements of the previously defined cities based on the state-of-the-art design rule and stochastic variations. In addition, the generated sewer systems are pipe-sized and exported as an SWMM (Rossman, 2004) input file.

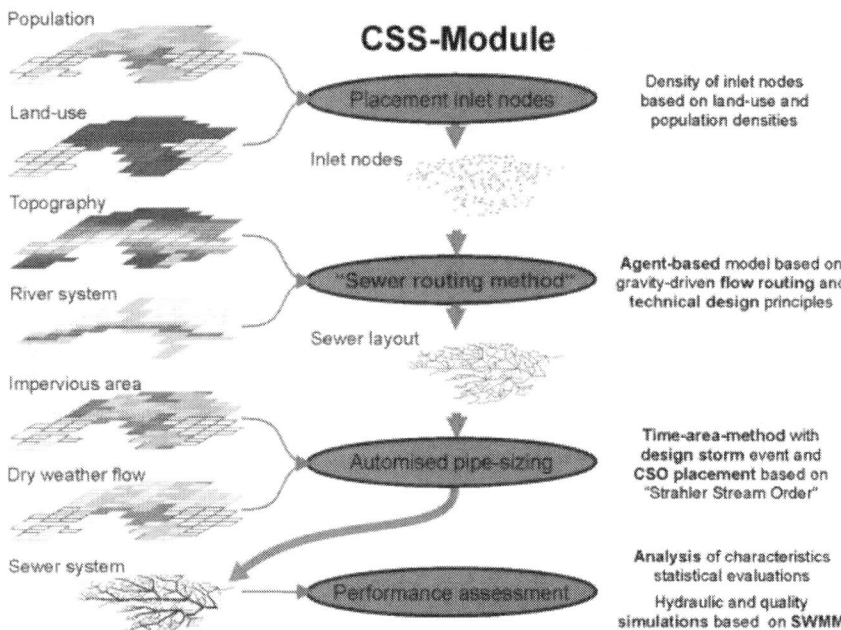

Figure 4. Procedural method of the CSS-Module.

The required inlet nodes to the sewer system are placed based on the GIS maps for population density and land use (Fig. 4, placement of inlet nodes). The inlet nodes of the sewer systems are automatically connected to a sewer layout, which drains the urban area to a waste water treatment plant (WWTP). To that, a "sewer routing method" has been developed which takes into account (gravity driven) flow routing and technical design principles for the layout of combined sewer systems (e.g., vertical alignment, WWTP, manhole spacing for direction changes, minimal slope and culverts for river-crossing). This creation process of the layout of the combined sewer system is applying agent-based modelling techniques. The movement paths of the agents represent the layout of the sewer pipes result in a fully branched network (see Fig. 4).

Subsequently, the sewer networks are algorithmically pipe-sized based on the time-area method (Butler and Davies, 2004) and a design-storm event (taking into account the spatial distributed data for impervious area and amount of dry weather flow; see Fig. 4). In these combined sewer networks, besides nodes and junctions, also storage units, weirs and combined sewer overflows are regarded. The resulting combined sewer systems are exported to the hydraulic solver SWMM (see Fig. 5). A detailed discussion of the entire process and its limitations can be obtained in Urich et al. (2010).

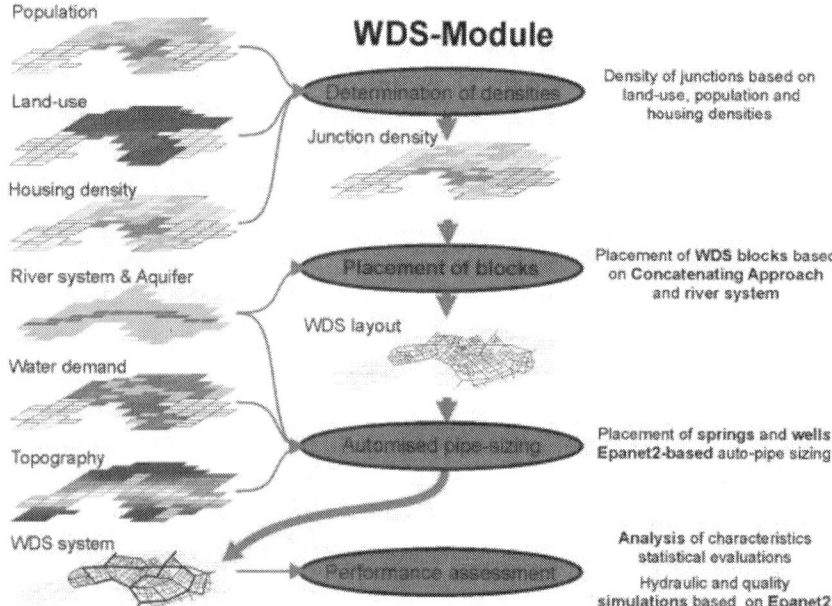

Figure 5. Procedural method of the WDS-Module.

Water distribution module (WDS-Module)
With the WDS-Module of VIBe (Sitzenfrei et al., 2010c), water distribution systems (WDS) are generated based on GIS data from the US-Module (see Figs. 3 and 5) or real-world data. The generation process is based on maps for topography, land use, population densities, housing densities, river systems, aquifers and water demand.

Based on the GIS maps for population density, land use and housing density, a map for the required junction densities is created (Fig. 5, placement of inlet nodes). The generation process of WDS implemented in VIBe follows the idea of complex network structures composed of recurring network motifs (Milo et al., 2002). Different network motifs (WDS-blocks) are designed by means of a graph theory based approach. The diverging WDS-blocks represent different redundancies and arrangements of pipe connections (e.g., branched or looped blocks) and different housing densities (assumed to correspond with network junction densities). For the generation process, a database of different WDS-blocks has been developed, tested and compared with real-world network structures (Sitzenfrei et al., 2013). With this database and the maps for housing and population densities, entire complex WDS can be composed by selecting WDS-blocks from the database that meet the requirements of the GIS data in the "city layout" and concatenating those WDS-blocks to the entire WDS. Subsequently, the water demand and elevation are added spatially distributed to junctions in the generated WDS. Finally, water sources are added to the generated WDS. Depending on the topographic and topological boundary conditions, hillside springs or groundwater wells are added. Therefore, gravity driven supply as well as pumped systems can be generated. Components for pressure management (like pressure reducing valves) are at the current state of development not regarded.

Next, the diameters of the WDS are designed based on the economic flow velocity. Using an iterative approach (Möderl et al., 2007), the flow velocities for the entire WDS are determined with the EPANET2 hydraulic solver. The diameters are then incrementally augmented as long as the actual flow velocity exceeds the defined economic one. For performance evaluation, an interface to the hydraulic solver EPANET2 is implemented.

INTEGRATED SCENARIO ANALYSIS OF WATER INFRASTRUCTURE

Fratini et al. (2012) noted that for the transition of traditional water infrastructure towards sustainability, it is important to be more engaged with other disciplines (e.g., social sciences, urban planning and architecture). The VIBe approach allows for an integrated scenario analysis that takes into account the entire urban water infrastructure and the structure of the city. The GIS information of the population and water demand is spatially linked to the water infrastructure models (WDS model and UD model). Therefore, changes in the population densities and the water demand can directly be linked to the WDS and UD models, respectively.

Coupling urban water infrastructure

The basis for all further calculations is the total daily average water demand, $Q_{d,m,t}$, determined with the spatial distribution of the population (population equivalents (PE)). From that, the relevant water flows and multiplication factors (i.e., prospective demand factor f_p, daily peaking factor $f_{d,max}$, hourly peaking factor $f_{h,max}$, maximum dry weather flow $f_{DWF,max}$) relevant for the design of systems of the investigated size (above 50,000 PE) and performance assessment (ÖNORM, 2002, ÖWAV-RB 11, 2009 and ÖWAV-RB 19, 1987) can be determined:

For the different design and performance assessment scenarios (see also Sections 3.2 and 4), the water flows in the hydraulic modelling processes are determined according to Table 1.

Table 1. Assessment and design flows for WDS and UD.

Description	Variable name	Factor(s)	Composition	Application (see also Section 3.2)
Total current daily average water demand	$Q_{d,m,t}$	–	–	Basis for further calculations
Water losses	Q_{WL}	$f_{WL} = 0.008\text{–}0.12$	$Q_{d,m,t} \times f_{WL}$	Fraction of $Q_{d,m}$
Current daily average water demand	$Q_{d,m}$	–	$Q_{d,m,t} - Q_{WL}$	WDS: current water quality performance
Future hourly peak flow of maximum day	$Q_{h,max,max}$, f	$f_p = 1.3$ $f_{d,max} = 1.4$ $f_{h,max} = 1.44$	$Q_{d,m} \times f_p \times f_{d,max} \times f_{h,max} + Q_{WL}$	WDS: network design
Current hourly peak flow of maximum day demand	$Q_{h,max,max}$	$f_{d,max} = 1.4$ $f_{h,max} = 1.44$	$Q_{d,m} \times f_{d,max} \times f_{h,max} + Q_{WL}$	WDS: current hydraulic performance
Sewer infiltration water	Q_i		0.75–1.25 l/(1000 PE d)	UD: fraction of $Q_{DWF,hyd}$
Current DWF for hydraulic performance (maximum DWF)	$Q_{DWF,hyd}$	$f_{DWF,max} = 1.2$	$Q_{d,m} \times f_{DWF,max} + Q_i$	UD: with rain weather flow for design and CSO and flooding assessment
Current DWF for shear stress performance (minimum DWF)	$Q_{DWF,\tau}$		120–150 l/(PE d)	UD: without rain for shear stress performance

Performance assessment

For the performance assessment, the generated networks are simulated with the software tools EPANET2 (water distribution system) and SWMM5 (urban drainage) by applying the water demands and dry weather flows according to Table 1. In the WDS model, the pressure distribution in hydraulic peak flow situations is assessed by means of a steady state simulation (only one time step). For the water quality analysis, the residence time of the water in the pipe network is investigated. Therefore, a simulation time of 50 h is used (series of steady state simulations).

For the UD model, a design-storm event of type EULER II (ATV-A 118E, 2006) with a return period of 5 years and a duration of 2 h is used. The simulation time for the rain weather simulation is 24 h.

For the performance assessment, the following performance indicators normalised between 0 and 1 are used:

- CSO performance (UD CSO): ratio between the total volume of the surface runoff that is treated at the waste water treatment plant and the total surface runoff.
- Flooding performance (UD flooding): 1 minus the maximum ponded volume over all nodes divided by the total rainfall runoff.
- Bed shear stress performance (UD shear stress): the bed shear stress is calculated according to the Austrian standard ÖWAV-RB 11 (2009) for each pipe i with Equation 1. A bed shear stress of minimum 1 N/m² indicates sufficient performance. Therefore, all pipes with a stress value above or equal to 1 N/m² are assessed as 1. For pipes below that threshold, the actual bed shear stress (hence between 0 and 1) is used for evaluation. The sum of all values (between 0 and 1) divided by the number of pipes results in the performance indicator used for the bed shear stress. In simplified terms, this performance indicator not only expresses the occurrence of sewer sedimentation but is representative for all issues related to low flow conditions, such as sewer gas production or corrosion.

$$\tau_i = \rho \cdot g \cdot S_i \cdot D_{H,p,i} (N/m^2) \qquad (1)$$

τi, shear stress for pipe i (N/m²); ρ, density of water 1000 (kg/m³); g, gravitational acceleration 9.81 (m/s²); Si, slope of pipe i; DH_p,i, hydraulic radius for partial filling for pipe i (m).

•Hydraulic performance (WDS hydraulic): for the hydraulic performance evaluation in the WDS model, the pressures in the junctions above or below the specified thresholds for Alpine systems (40 or 100 m, respectively) are assessed as 0. Between these two thresholds, a junction is assessed as 1. The sum over all junctions weighted with the demand divided by the number of junctions and the total demand results in the performance indicator for the hydraulic performance in the WDS.

- Water quality performance (WDS quality): for water quality, the water age (in this study, the travel time in the pipe network) is used as the indicator. Bacterial growth or other quality issues can be related to that indicator. If the travel time from the source to the demand nodes is lower than the threshold of 24 h, it is assessed as 1. If it is above this threshold, it is assessed as 0. The sum over all junctions weighted with the demand divided by the number of junctions and the total demand results in the performance indicator for the quality performance.

- Integrated performance: for an integrated performance assessment of both the water distribution and the urban drainage system, an overall performance as a product of all the single performance indicators described above is introduced.

For each performance indicator, a value of 1 indicates excellent performance and a value of 0 indicates total failure of the system (in terms of the used threshold values).

DEFINITIONS OF SCENARIOS AND CASE STUDIES

With the presented VIBe approach, 80 virtual Alpine cities (populations between 70,000 and 170,000) including the water distribution system and urban drainage system are generated. In addition, the real-world city Innsbruck in Austria (population 121,000) is analysed. For each of the generated city-scale case studies, different scenarios for changes in the population, water demand and amount of produced dry weather flow are investigated based on hydraulic simulations. These changes can represent different transition scenarios from centralised to decentralised water infrastructure. Although a time dependent change of the layout of water networks (e.g., network expansion, rehabilitation measures) is neglected, different levels of decentralisation (scenarios defined in the following) can represent such a transition over time. Such changes can be very fast (implementation in a few years), but can also be applicable to represent stepwise changes over a longer time period. For the latter, it must be assumed that the layout of the water networks remain static over the time horizon. Despite this limitation, the procedure allows to assess to what extent a transition to decentralised water solutions can be implemented before the existing centralised systems are pushed to their limit.

In this study, the assumption is made that all changes are uniformly distributed in the entire city. Although reductions may concentrate in specific areas and therefore cause specific local problems in real-world cases, this aspect is no regarded here for simplicity. However, disregarding such aspects does not change the presented approach. More detailed information about the change in the population and land use could be obtained and implemented by applying an urban development model to determine possible future conditions (Sitzenfrei et al., 2010b). To investigate possible transitions and unknown future developments systematically, the following scenarios are defined:

- Reduction scenarios: decrease of the population and reduction in the potable water demand per capita due to, for instance, water saving measures, decentralised water supply and reuse. Therefore, reduction rates can be applied to the daily average potable water demand, while water losses are kept constant (between 8% and 12% of the average potable water demand). This also impacts the produced dry weather flow, and the impervious area is assumed to remain as is. The five defined scenarios are a reduction to 80%, 60%, 40%, 20% and 10% of the initial water demand and dry weather flow production.
- Increase scenarios: increase of the population and increase in the water demand per capita. This increase is applied to the water demand and dry weather flow production, respectively, while the water losses are kept constant. In addition, a linear regression between the change in the population density and imperviousness is assumed (Chabaeva et al., 2004; impervious area (%) = population density ($m^2 \times 10^4$) × 0.492 + 16.732). The five defined scenarios are increases of 20%, 40%, 60%, 80% and 100% of the initial water demand and increase of the impervious area calculated with the linear regression mentioned above.

Including the assessment of the initial situation, 11 variations for the performance assessment are evaluated for each of the 80 virtual case studies and the real-world case study Innsbruck. This results in a total of 891 investigated city-scale scenarios.

Case study water infrastructure Innsbruck
The Alpine city has approximately 121,000 inhabitants. The centralised potable water system is supplied mainly by one hillside spring, and water treatment is not required. The height difference between the source and most of the supply area is 100 m or more. Therefore, the regular supply is completely driven by gravity. For emergency supply, groundwater wells are installed. The average annual water demand is approximately 11.7 million m^3, and the hourly peak demand on a maximum day is approximately 1000 l/s. The hydraulic model (EPANET2) consists of 7128 junctions and 7599 pipes.

A combined sewer system drains the urban area to a centralised waste water treatment plant. In the UD system of Innsbruck, dry weather inflow from outlying conurbation areas is fed in. To determine the spatial distribution of the population for dry weather simulations, the approximate spatial distribution derived from the water demand (WDS model) is used.

RESULTS

The hydraulic model (SWMM5) consists of 182 subcatchments, 247 junction nodes and 275 conduit links.

RESULTS

Integrated generation process – VIBe city

In Fig. 6, an example virtual city is visualised including the GIS data for population density, dry weather flow production, impervious area and land use classes. The hydraulic models of the water infrastructures are also shown.

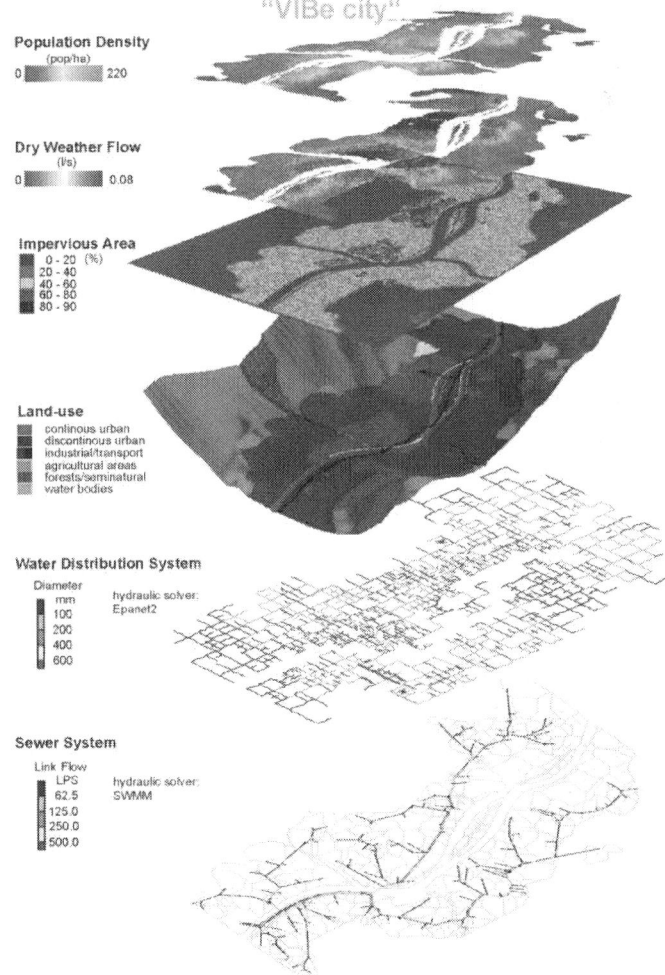

Figure 6. City generated with VIBe.

In Fig. 7(a), the cumulative distribution function of the population of the 80 generated Alpine cities is shown. The size of Innsbruck is approximately the median value of the generated systems ($F(x) = 0.51$). Three size classes are defined: (1) population below 100,000 (minimum 70,000) – class 1, about a fourth of the systems; (2) population between 100,000 and 140,000 – class 2, about half of the systems including Innsbruck; and (3) population over 140,000 (maximum 170,000) – class 3, about a fourth of the systems.

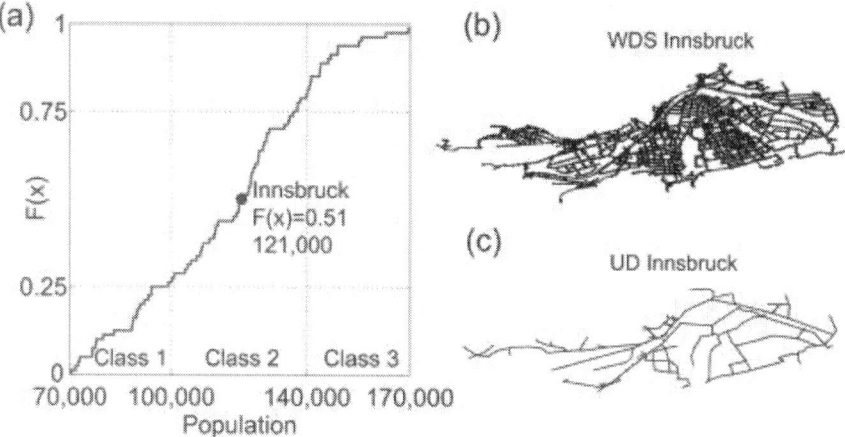

Figure 7. Cumulative distribution function of the generated VIBe cities with class definition (a), water distribution (b) and drainage system (c) of the real case study Innsbruck.

Impact of transitions from centralised to decentralised water infrastructures

Fig. 8 shows the impact of the reduction and increase scenarios (expressed as variation factors) on the performance of the Innsbruck water networks. The graph on the left inFig. 8 shows the impact for the bed shear stress as a cumulative distribution function for all pipes, and the graph on the right in Fig. 8 shows the network travel time in the WDS model. The shear stress evaluations are based on the water height calculations with SWMM5 for dry weather flow. The network travel time is a result of the EPANET2 simulations. The thick black line shows the baseline values. The dashed grey line shows the threshold values used for the performance evaluations (for the performance indicators, see Section 3.2).

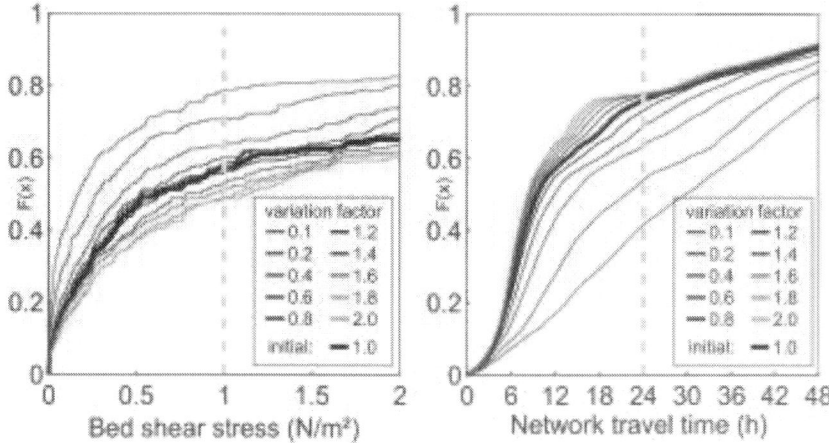

Figure 8. Cumulative distribution function and the impact of the variation factors on the shear stresses in the urban drainage system of Innsbruck (left) and the water age in the water distribution system of Innsbruck (right).

For the bed shear stress, it can be observed that almost 60% of the values are below the threshold of 1 N/m². An increase in the dry weather flow production does not change this picture substantially, whereas a decrease of the variation factor to 0.1 raises the percentage of values below 1 N/m² to approximately 80%. For the network travel time in the WDS model, approximately 75% of the water ages in the junctions are below the threshold value of 24 h for the initial system. An increase in the water demand has less impact (shift in percentages $F(x)$) than a decrease.

In Fig. 9, the evaluations of all generated systems are shown aggregated to the normalised five performance indicators (five columns: UD CSO, UD flooding, UD shear stress, WDS hydraulic, WDS quality) described before. For each size class (three rows: Class 1–Class 3), a row of boxplots is shown. In each of the 15 boxes, there are 11 boxplots (for the initial, the increase and the reduction scenarios). The dots in the circles indicate the median values of the virtual systems in that class for the performance indicator and the particular variation. In addition, the performance of the real-world case study Innsbruck is shown with a circle in each boxplot.

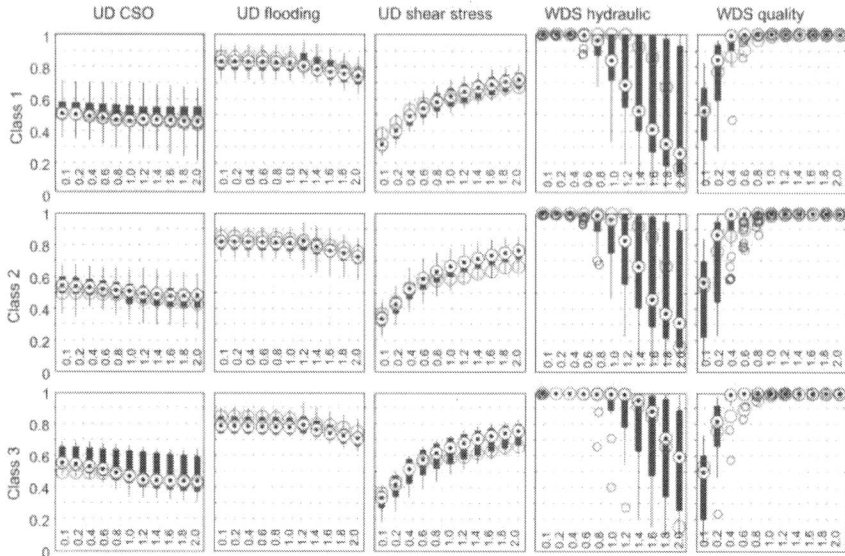

Figure 9. Boxplots of the performance indicators for the three different size classes.

Stability of the water networks

For the UD CSO performance, reductions in the dry weather flow have only a marginal impact on the performance. Thus, with regard to the transition from centralised to decentralised urban water solutions, the impact on the combined sewer system is marginal in terms of the CSO efficiency because this is mainly driven by rain weather flow. For 75% of the investigated systems, the maximum reductions in the CSO efficiency are smaller than −10%. For the case study Innsbruck, the maximum reduction is −10.9%.

In terms of the UD flooding performance, a dry weather flow reduction also only has a marginal impact on the performance. Again, per definition, this performance is mainly driven by rain weather flow. Thus, for a dry weather increase and for additional impervious area, the flooding efficiency decreases. However, the maximum reduction is more than −20% for approximately 5% of the systems which indicates that for most of the investigated systems (95%) the maximum reduction is less than 20%. The flooding efficiency and CSO efficiency are inversely proportional.

Only the bed shear stress performance is affected significantly by dry weather flow variations. A reduction of the dry weather flow by a factor smaller than 0.6 decreases the performance. In general, an increase produces only marginal improvements. The water demand reduction usually has no impact on the hydraulic pressure performance (WDS hydraulic). Conversely, an increase in the water demand leads to a pressure reduction because of increased friction losses. For the water quality, adequate performance can be assured even for a water demand of 60% provided that tank management is sufficient (filling and emptying cycles of the tanks), and that the demand reduction is equally distributed.

The WDS hydraulic performance (pressure performance) for small systems (class 1) is affected more compared to medium and large systems (classes 2 and 3). In particular, the small and medium systems reach their hydraulic limits earlier when increasing demands. In other words, the large systems are for the investigated scenarios more stable in terms of the demand increase because of their larger and redundant capacities. For the case study Innsbruck, a rapid drop in the hydraulic pressure performance is observed for increase factors above 1.4. This reduction is observed because the water resources in Innsbruck are (although available) not sufficiently exploited for such a high amount of water demand. The generated systems can have more capacity or can cover more demand because of the implemented simple design procedure which is solely based on economic flow velocities (for details see Sitzenfrei et al., 2013). For all classes, there is no impact in terms of the hydraulic performance when reducing the water demand. The water quality performance does not show any effects for water demand increases. For a water demand decrease, water quality reductions are visible as soon as the demand is reduced to 80% or less. In terms of transitions to, for instance, decentralised water recycling or water savings, the centralised urban drainage systems and water distribution system are sufficient until a reduction of −40% of the design water demand and dry weather flow, respectively.

Fig. 10 shows the integrated performance (multiplication of all performance indicators) against the different variation factors. The different performance indicators might be weighted differently, but for simplicity and to show the application, in this work they are weighted equally. The maximum values for classes 1 and 2 are comparable (95% percentile curve), but the 75% and 50% curves exhibit different behaviour. Comparing the different classes indicates that the small systems (class 1) are more likely to exhibit a lower integrated performance and to be (for the investigated scenarios) less stable in terms of the variations.

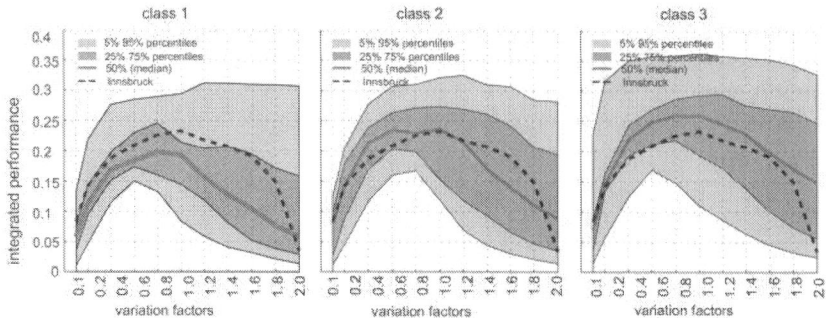

Figure 10. Integrated performance – stability of the networks.

In class 2, only approximately 25% of the generated systems (over 75 percentile) are negligibly affected (variation ranges of 0.6 and 1.4). Although the case study Innsbruck has its best overall performance for the initial state (variation factor 1.0), the system is redundant and is not significantly affected for variation factors between 0.4 and 1.6. The large systems (class 3) are in general more likely to respond with a stable performance because of the applied variation factors.

MODEL BUILDING EXAMPLE WITH NUMEROUS VIRTUAL CASE STUDIES

To note the benefit of using numerous virtual case studies for model building, the changes in the shear stress performance are discussed and interpreted in more detail. For the system performance indicator of the shear stress efficiency (UD shear stress), different variables, such as the hydraulic radius for different diameters and shapes and different slopes (see Equation 1), are taken into account. Likewise, the efficiency is influenced by the topology of the drainage network (e.g., network loops). However, the change in that performance indicator can be traced back to the change in the total dry weather flow production and therefore to the applied variation factors. In the graph on the left in Fig. 11, a base 10 logarithm is applied to all variation factors of the investigated systems. The 1, 5, 95 and 99 percentiles as well as the median value are shown. An almost linear coherence can be observed in that semi logarithmic plot.

MODEL BUILDING EXAMPLE WITH NUMEROUS VIRTUAL CASE STUDIES

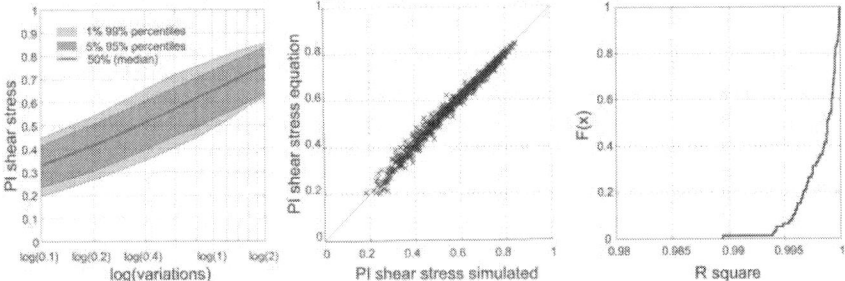

Figure 11. *PI* shear stress for the logarithm of the variations (left); regressions between the *PI* shear stress equation and simulation (middle); R^2 for regressions from different starting points (right).

In Equation 2, the regression is formulated for the observed data. With an observed or simulated performance indicator for the shear stress ($PI_{S,i}$) and a corresponding total water flow in the combined urban drainage system (Q_i), the performance indicator ($PI_{S,n}$) for a changed dry weather flow production (Q_n) can be predicted with Equation 2:

$$PI_{S,n} = PI_{S,i} + \log_{10}\frac{Q_n}{Q_i}0.3227 \text{ and } Q_n/Q_i \in [0.05, 20] \quad (2)$$

$PI_{S,n}$ calculated performance indicator for the shear stress for a total dry weather flow production in the system for Q_n (–); $PI_{S,n} \in [0,1]$, $Q_n \neq 0$, $Q_i \neq 0$,

$PI_{S,i}$ assessed performance indicator for the shear stress for a total dry weather flow production in the system for Q_i (–); $PI_{S,i} \in [0,1]$, $Q_i \neq 0$, $Q_i \neq 0$.

For example, with an initial shear stress performance indicator of 0.7 (obtained value for the current situation $PI_{S,i}$), a bisection of dry weather flow ($Q_n/Q_i = 0.5$) reduces the shear stress performance to 0.6 (0.7 + log 10(0.5) × 0.3227). In the middle figure in Fig. 11, the results obtained with Equation 2 applied to the investigated systems are compared to the simulated results. For each of the 80 virtual case studies, in total 10 scenarios are tested (one marker represents such a scenario, resulting in 800 data points). Separately for each virtual case study, a regression analysis was performed.

In the figure on the right in Fig. 11, the cumulative distribution function of these 80 results for R^2 is plotted. The median value of R^2 for the 80 virtual cities is 0.998. All R^2 values are above 0.988. Comparing the results from

Equation 2 to the simulation results of the case study Innsbruck gives an R^2 of 0.9904.

Solely with the results of the case study Innsbruck, Equation 2 could also have been developed. Although the results for the case study Innsbruck are also promising, Equation 2 would be very case specific and therefore less convincing as the analysis shown above.

CONCLUSIONS

The transition of traditional urban water systems towards more sustainable solutions has significant effects on the remaining central water networks. For a comprehensive assessment of the impact, it is necessary to investigate a large number of case studies. However, for modelling real-world systems, this is a very difficult undertaking because data collection, model building and calibration are tedious. In this paper, an alternative approach is presented and applied based on the stochastic generation of virtual case studies. VIBe (Virtual Infrastructure Benchmarking) is a tool that enables the stochastic generation of urban water systems for case study research with implemented data interfaces of external modelling software. An analysis of transition scenarios for water infrastructure based on these numerous case study data with varying characteristics leads to general, case-unspecific conclusions. In this work, an analysis of 80 virtual and one real-world case study is used to compare performance in systems of different sizes. Investigating the impact of the water demand decrease (e.g., due to water saving measures and transition in the implementation of decentralised water reuse systems) and water demand increase (e.g., population increase) on the water network reveals if adequate hydraulics and quality can be maintained in the network. Investigations related to numerous virtual systems reveal if the obtained results for the real-world system are outliers or a regular response to the investigated changes. Thus, percentages are determined to study how likely a system with a specific characteristic (e.g., population size) is to demonstrate representative behaviour. The percentages are determined based on evaluations of percentiles of performance assessments of many different systems. It is revealed that small systems (in context of this study defined as an Alpine system with a population between 70,000 and 100,000) are more affected as compared to medium (100,000–140,000) and large systems (140,000–170,000). In particular, large systems are for the investigated scenarios more stable in terms of the demand increase because of their larger and redundant capacities. Although the maximum values for small and medium systems are comparable, 75% of the small systems were identified to have a less stable integrated performance. A quarter of the medium systems are negligibly affected for variation ranges between 0.6 and 1.4. The investigated changes can represent transition scenarios over time. E.g., for the first decentralisation step there is a reduction factor of 0.8 applied and in second 0.6 can be regarded. This resembles a quasi-stationary consideration of dynamics. A time dependent change in the layout of the water networks (e.g., rehabilitation measures or network expansion) was not regarded here. Therefore, a representation of stepwise transition scenarios over a longer time horizon in the context of this paper is only applicable, as long as neglecting time dependent changes of the networks is appropriate.

Based on the analysis of 80 virtual and one real-world case study, a simple equation was derived to estimate the change in the shear stress performance due to changes in the dry weather flow production (water demand change). For the investigated systems, excellent agreements were obtained by comparing the complex, simulated and simplified, determined performances. Therefore, with the developed simplified approach, the shear stress performance can be predicted for dry weather flow changes for systems.

ACKNOWLEDGEMENTS

This work was funded by the Austrian Science Fund (FWF) in the project DynaVIBeP23250-N24. The authors gratefully acknowledge the financial support.

REFERENCES

1. Achleitner, S., Mo¨derl, M., Rauch, W., 2007. Urine separation as part of a real-time control strategy. Urban Water J. 4 (4), 233e240.
2. ATV-A 118E, 2006. Hydraulic Dimensioning and Verification of Drainage Systems. ATV e.V., Hennef. Barton, A.B., Argue, J.R., 2009. Integrated urban water management for residential areas: a reuse model. Water Sci. Technol. 60 (3), 813e823.
3. Blumensaat, F., Wolfram, M., Krebs, P., 2012. Sewer model development under minimum data requirements. Environ. Earth Sci. 65 (5), 1427e1437.

4. Borsanyi, P., Benedetti, L., Dirckx, G., De Keyser, W., Muschalla, D., Solvi, A.M., Vandenberghe, V., Weyand, M., Vanrolleghem, P.A., 2008. Modelling real-time control options on virtual sewer systems. J. Environ. Eng. Sci. 7 (4), 395e410.
5. Brown, R.R., Keath, N., Wong, T.H.F., 2009. Urban water management in cities: historical, current and future regimes. Water Sci. Technol. 59 (5), 847e855.
6. Butler, D., Davies, J.W., 2004. Urban Drainage, second ed. Spon Press, London, ISBN 0-415-30607-8. Butler, D., Schu¨tze, M., 2005. Integrating simulation models with a view to optimal control of urban wastewater systems. Environ. Model. Software 20 (4), 415e426.
7. Chabaeva, A., Civco, D., Prisloe, S., 2004. Development of a population density and land use based regression model to calculate the amount of imperviousness. In: Proceedings of the ASPRS Annual Convention. Denver. De Toffol, S., Engelhard, C., Rauch, W., 2007.
8. Combined sewer system versus separate system e a comparison of ecological and economical performance indicators. Water Sci. Technol. 55 (4), 255e264.
9. Dome`nech, L., Saurı´, D., 2010. Socio-technical transitions in water scarcity contexts: public acceptance of greywater reuse technologies in the Metropolitan Area of Barcelona. Resour. Conserv. Recy. 55 (1), 53e62.
10. Dong, X., Zeng, S., Chen, J., 2012. A spatial multi-objective optimization model for sustainable urban wastewater system layout planning. Water Sci. Technol. 66 (2), 267e274.
11. Fratini, C.F., Elle, M., Jensen, M.B., Mikkelsen, P.S., 2012. A conceptual framework for addressing complexity and unfolding transition dynamics when developing sustainable adaptation strategies in urban water management. Water Sci. Technol. 66 (11), 2393e2401.
12. Girona´s, J., Roesner, L.A., Rossman, L.A., Davis, J., 2010. A new applications manual for the Storm Water Management Model (SWMM). Environ. Model. Software 25 (6), 813e814. Lau, J., Butler, D., Schu¨tze, M., 2002.
13. Is combined sewer overflow spill frequency/volume a good indicator of receiving water quality impact? Urban Water 4 (2), 181e189.
14. Le Coustumer, S., Fletcher, T.D., Deletic, A., Barraud, S., Poelsma, P., 2012. The influence of design parameters on clogging of storm water biofilters: a large-scale column study. Water Res. 46 (20), 6743e6752.
15. Meirlaen, J., Huyghebaert, B., Sforzi, F., Benedetti, L., Vanrolleghem, P., 2001. Fast, simultaneous simulation of the integrated urban wastewater system using mechanistic surrogate models. Water Sci. Technol. 43 (7), 301e309.

16. Milo, R., Shen-Orr, S., Itzkovitz, S., Kashtan, N., Chklovskii, D., Alon, U., 2002. Network motifs: simple building blocks of complex networks. Science 298 (5594), 824e827. Mo"derl, M., Butler, D., Rauch, W., 2009.
17. A stochastic approach for automatic generation of urban drainage systems. Water Sci. Technol. 59 (6), 1137e1143. Mo"derl, M., Fetz, T., Rauch, W., 2007. Stochastic approach for performance evaluation regarding water distribution systems. Water Sci. Technol. 56 (9), 29e36.
18. Mo"derl, M., Sitzenfrei, R., Fetz, T., Fleischhacker, E., Rauch, W., 2011. Systematic generation of virtual networks for water supply. Water Resour. Res. 47.
19. Moore, S.L., Stovin, V.R., Wall, M., Ashley, R.M., 2012. A GIS-based methodology for selecting storm water disconnection opportunities. Water Sci. Technol. 66 (2), 275e283.
20. Ole, F., Torben, D., Bergen, J.M., 2012. A planning framework for sustainable urban drainage systems. Water Policy 14 (5), 865e886.
21. O" NORM B 2538, 2002. Transport-, Versorgungs- und Anschlussleitungen von Wasserversorgungsanlagen e Erga"nzende Bestimmungen zu O" NORM EN 805.
22. O" sterreichisches Normungsinstitut, Wien. O" WAV-RB 11, 2009. Richtlinie fu" r die abwassertechnische Berechnung und Dimensionierung von Abwasserkana"len. O" sterreichisches Normungsinstitut, Wien.
23. O" WAV-RB 19, 1987. Richtlinie fu" r die Bemessung und Gestaltung von Regenentlastungen in Mischwasserkana"len (Guideline for design and construction of combined sewer overflows).
24. O" sterreichisches Normungsinstitut, Wien. Peter-Varbanets, M., Zurbru" gg, C., Swartz, C., Pronk, W., 2009. Decentralized systems for potable water and the potential of membrane technology. Water Res. 43 (2), 245e265.
25. Rauch, W., Brockmann, D., Peters, I., Larsen, T.A., Gujer, W., 2003. Combining urine separation with waste design: an analysis using a stochastic model for urine production. Water Res. 37 (3), 681e689.
26. Rauch, W., Harremoe"s, P., 1999. On the potential of genetic algorithms in urban drainage modeling. Urban Water 1 (1), 79e89.
27. Refsgaard, J.C., Henriksen, H.J., 2004. Modelling guidelines e terminology and guiding principles. Adv. Water Resour. 27 (1), 71e82.
28. Rossman, L.A., 2000. EPANET 2 User Manual. National Risk Management Research Laboratory e U.S. Environmental Protection Agency. Rossman, L.A., 2004.
29. Storm Water Management Model e User's Manual Version 5.0. National Risk Management Research Laboratory e U.S. Environmental Protection Agency. Scheidegger, A., Hug, T., Rieckermann, J., Maurer, M., 2011. Network

condition simulator for benchmarking sewer deterioration models. Water Res. 45 (16), 4983e4994.
30. Scheidegger, A., Maurer, M., 2012. Identifying biases in deterioration models using synthetic sewer data. Water Sci. Technol. 66 (11), 2363e2369.
31. Schilling, W., 1989. Real Time Control of Urban Drainage Systems. The State-of-the-Art, IAWPRC Scientific and Technical, Report N. 2. Schu¨tze, M., Butler, D., Beck, M.B., 1999. Optimisation of control strategies for the urban wastewater system e an integrated approach. Water Sci. Technol. 39 (9), 209e216.
32. Schu¨tze, M., Butler, D., Beck, M.B., Verworn, H.R., 2002. Criteria for assessment of the operational potential of the urban wastewater system. Water Sci. Technol. 45 (3), 141e148. Sitzenfrei, R., 2010.
33. Stochastic Generation of Urban Water Systems for Case Study Analysis. University of Innsbruck, Unit of Environmental Engineering. Sitzenfrei, R., Fach, S., Kinzel, H., Rauch, W., 2010a. A multi-layer cellular automata approach for algorithmic generation of virtual case studies e VIBe. Water Sci. Technol. 61 (1), 37e45.
34. Sitzenfrei, R., Fach, S., Kleidorfer, M., Urich, C., Rauch, W., 2010b. Dynamic virtual infrastructure benchmarking: DynaVIBe. Water Sci. Technol.: Water Supply 10 (4).
35. Sitzenfrei, R., Mo¨derl, M., Rauch, W., 2010c. Graph-based approach for generating virtual water distribution systems in the software VIBe. Water Sci. Technol.: Water Supply 10 (6), 923e932.
36. Sitzenfrei, R., Mo¨derl, M., Rauch, W., 2013. Automatic generation of water distribution systems based on GIS data. Environ. Model. Software 47, 138e147.
37. Trifunovic, N., Maharhan, B., Vairavamoorthy, K., 2012. Spatial network generation tool for water distribution network design and performance analysis. Water Sci. Technol.: Water Supply 13 (1).
38. Urich, C., Sitzenfrei, R., Moderl, M., Rauch, W., 2010. An agentbased approach for generating virtual sewer systems. Water Sci. Technol. 62 (5), 1090e1097.
39. Ward, S., Memon, F.A., Butler, D., 2012. Performance of a large building rainwater harvesting system. Water Res. 46 (16), 5127e5134.
40. Zacharof, A.I., Butler, D., Schu¨tze, M., Beck, M.B., 2004. Screening for real-time control potential of urban wastewater systems. J. Hydrol. 299 (3e4), 349e362.

CITATION

Robert Sitzenfrei, Michael Möderl, Wolfgang Rauch, Assessing the impact of transitions from centralised to decentralised water solutions on existing infrastructures – Integrated city-scale analysis with VIBe, Water Research, Volume 47, Issue 20, 15 December 2013, Pages 7251-7263, ISSN 0043-1354, http://dx.doi.org/10.1016/j.watres.2013.10.038.

CHAPTER 2

Detection of Water Pipes and Leakages in Rural Water Supply Networks Using Remote Sensing Techniques

Diofantos G. Hadjimitsis[1], Athos Agapiou[1], Kyriacos Themistocleous[1], Dimitrios D. Alexakis[1], Giorgos Toulios[1], Skevi Perdikou[2], Apostolos Sarris[3], Leonidas Toulios[4] and Chris Clayton[5]

[1] Cyprus University of Technology, Faculty of Engineering and Technology, Department of Civil Engineering and Geomatics, Remote Sensing and Geo-Environment Lab, Cyprus

[2] Frederick University, Cyprus

[3] Laboratory of Geophysical, Satellite Remote Sensing and Archaeoenvironment, Institute for Mediterranean Studies, Foundation for Research and Technology, Hellas (F.O.R.T.H.), Cyprus

[4] Hellenic Agricultural Organisation DEMETER (NAGREF), Institute of Soil Mapping and Classification, Larissa, Greece

[5] University of Southampton, UK

INTRODUCTION

Water leakages have been a major problem for many regions around the world (Weifeng et al. 2011). However, monitoring such leakages is a difficult task since traditional field survey methods are costly and time consuming (Huang et al. 2010). Researchers from diverse scientific fields have studied this problem through the development of several techniques including radar technique, geophones, gas filling, and many others. Different conventional techniques such as acoustics, radioactive, electromagnetic, ground penetrating radar and linear polarization resistance have been used over the years for water pipeline leakage detection (Skolnik, 1990; Heathcote and Nicholas, 1998; Hunaidi and

Giamou, 1998; Eyuboglu *et al.*, 2003; Burn *et al.*, 2001; Hadjimitsis, *et al.*, 2009).

Remote sensing has been used for a wide range of applications including water management. Studies have shown promising results from its use for water leakage detection (Sheikh Naimullah, 2007). The uses of remote sensing techniques for water leakage detection are time and cost effective compared with traditional, intrusive methods, but their use is restricted due to their spatial resolution. The pipeline leakages occur along the length of the pipeline and the area affected may not be detectable by the satellite sensor as it depends on the pixel size and the density of the vegetation developed due to the presence of water.

Vegetation indices (VI) are the main form of satellite spectral data used for several applications. According to Agapiou *et al.* (2012a), VIs can be divided into five main categories according to equation or the use of each index, which include broadband indices, narrowband indices (hyperspectral), leaf pigment indices, stress indices and water stress indices. They reported that VI can be simply divided according to the wavelength characteristics used in their formula (broadband and narrowband indices). Using airborne remotely sensed imagery, Pickerill and Malthus (1998) analyzed two known water leaks and found that different vegetation indices and single bands were required in order to identify each leak. The spectral profile of one leak responded best to a ratio of NIR to red reflectance, while in the other, NIR to red reflectance ratio was not useful in differentiating it from its surroundings.

Huang et al. (2009) used airborne multispectral remote sensing imagery with high-resolution imaging sensors in the visible, NIR and thermal infrared wavelengths and found that airborne multispectral imaging is a useful tool in the detection of irrigation canal leakage in distribution networks. They concluded that the analysis of the processed image data from red, NIR and thermal bands is highly consistent with the observations from field investigation. Images from individual bands, particularly from the thermal band, can help detect leakage from irrigation canals. The NDVI image, which combines the data from the red and the NIR bands, can help detect and correct errors observed on the thermal imagery.

On-site observation, which consists of data collection, periodical observations, and multivariate risk assessment analysis, is the most common technique of monitoring the water pipe network in Cyprus. However, this is difficult to accomplish with traditional methods since it is time consuming, expensive and monitoring is localized. Furthermore, part of the water network tends to be located in inaccessible areas, away from the main road network and urban areas. A complete geoinformation system providing the exact location, characteristics and relevant data for the water mains does not exist, making the leakage monitoring procedures even more challenging.

This paper presents the results from a project which combines different remote sensing technologies for the detection and monitoring of water leakages for water utility systems located in open fields in Cyprus. Two case studies areas were evaluated using freely distributed Landsat 7 ETM+ satellite images and ground spectroradiometric data. In addition, a low altitude system was deployed to observe these pipelines from different heights.

Finally, different remote sensing techniques have been used evaluated as in the detection of leakage from a major water pipe in Cyprus ("*Southern Conveyor Project*"). Although significant efforts have been made to detect possible water leakages, as shown above, the detection of the water pipe itself it still problematic . This is because such water pipes networks are commonly mapped in a digital form (e.g. GIS environment). However, in most cases the digital location of the water pipe does not fully correspond with the real world, since many obstacles during the construction can be arise and therefore the route of the proposed pipe can change.

In order to explore further the capabilities of remote sensing –beyond the detection of water leakages- the authors have applied several algorithms for the detection of buried water pipes. The detection of buried features is well established procedure in archaeological research since buried anthropogenic remains can be found using remote sensing techniques (Agapiou *et al.*, 2010, 2012b; Sarris *et al.*, 2013). Indeed, soil marks or crop marks related with water pipes can be used, in a similar approach, for mapping the real footprint of a pipe network.

STUDY AREAS

In this section, three different case studies are presented. In the first case study, a part of the *"Southern Conveyor Project"* is described; following, two case studies for the *"Lakatameia"* and the *"Choirokoitia -Frenaros"* water pipes are presented. In the first case study, the authors have focused to the detection of the actual footprint of the pipe while in the next two case studies, remote sensing techniques have been evaluated for the detection of water leakages. The *"Lakatameia"*is a pipeline which is currently not in use while the *"Choirokoitia -Frenaros"* is a major pipeline of Cyprus where three major leakages have been recorded between 2007 to 2010.

"Southern Conveyor Project"

Water resources development in Cyprus initially focused on groundwater and, until 1970, groundwater was the main source of water supply for both drinking and irrigation purposes. As a result, almost all aquifers were seriously depleted because of over pumping. In addition, seawater intrusion was observed in most of the coastal aquifers. The increase of population as well as the increase in the tourist and industrial activities have led to an increase in the demand for water and have created an acute shortage of potable water.

Under these conditions, the implementation of the *"Southern Conveyor Project"* was a necessity and a basic prerequisite for the further agricultural and economic development of the island. The *"Southern Conveyor Project"* is the largest water development project ever undertaken by the Government of Cyprus. The basic objective of the project is to collect and store surplus water flowing to the sea and convey it to areas for both domestic water supply and irrigation. Essentially, the project aims to support the agricultural development of the coastal region between Limassol and Famagusta, as well as to meet the domestic water demand of Limassol, Larnaca, Famagusta, Nicosia, and a number of villages. In addition it supports the tourist and industrial demand of the southern, eastern and central areas of the island. The project is able to supply 33 million cubic metres of water for the irrigation of 13 926 hectares and another 33 million cubic metres of water for domestic purposes (Cyprus Water Development Department, 2000). In this case study, a part in the SE of Cyprus was examined (Figure 1).

STUDY AREAS

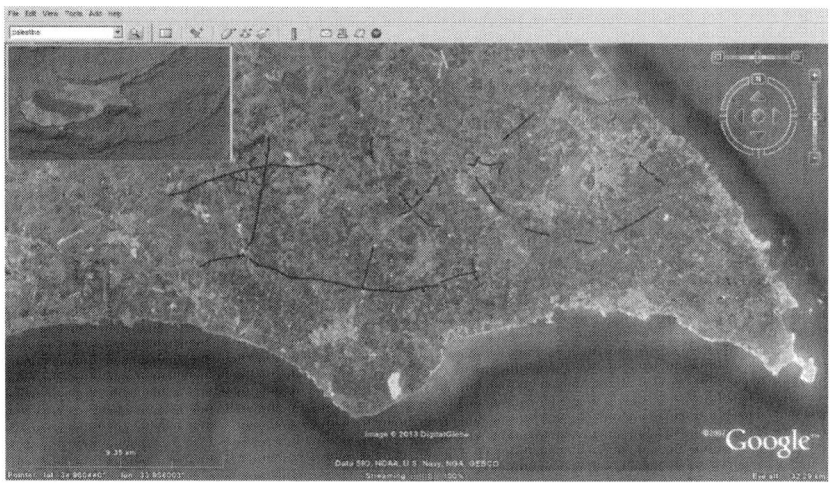

Figure 1. Map of the SE of Cyprus showing parts of "Southern Conveyor Project" (blue line) (© Google Earth)

"Lakatameia" Pipeline

An existing pipeline in the area of Lakatameia (central Cyprus) was selected to be used for the pilot study (Figure 2). The existing pipeline, with a length of less than 5 km, has been systematically reported as problematic due to several leakages and is therefore no longer in use by local authorities. The waterpipe runs through both urban and rural areas (see Figure 2). A section of the pipeline with a length of over 2km and located in a rural area, has been used to apply the different remote sensing techniques for the detection of leakages. Since the existing waterpipe is not currently used, it was necessary to fill the pipe with water periodically in order to observe the effectiveness of such remote sensing techniques.

The water pipe is made of UPVC and has 315mm diameter. It is between 1.80m and 2.00m below the ground surface and runs along the *Pediaos* river for a large part of its length. It is not currently being used due to water leakages occurring throughout almost the entire length of the pipeline. Information regarding the specific dates of the leakages is not available from local authorities.

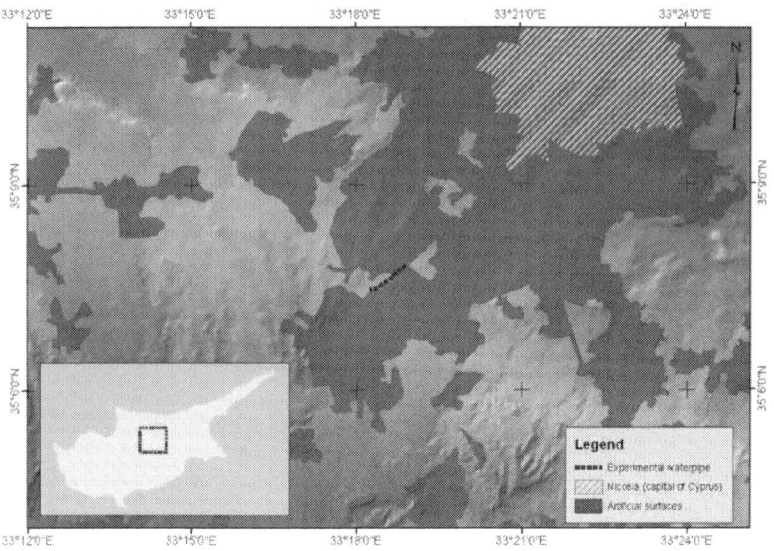

Figure 2. The *"Lakatameia"* waterpipe (dash line) used as the pilot study area.

"Frenaros — Choirokoitia" Water Pipe

The next area of interest is a major rural pipeline in Cyprus, which runs from the Choirokoitia area to the Frenaros area (Figure 3). The existing pipeline, which passes through the central and central-east part of Cyprus, has a length of over 65 km. The pipeline is located 1-3 meters below ground surface. Various geological formations, including calcaric cambisols, calcaric regosols, and epipetric calcisols exist in the area. elevation of the pipeline (ground surface) varies between 10 m and 200 m above sea level (Figure 4). In addition, the waterpipe passes through different types of land cover, as recorded from the CORINE 2000 land use map (Figure 5).

STUDY AREAS 35

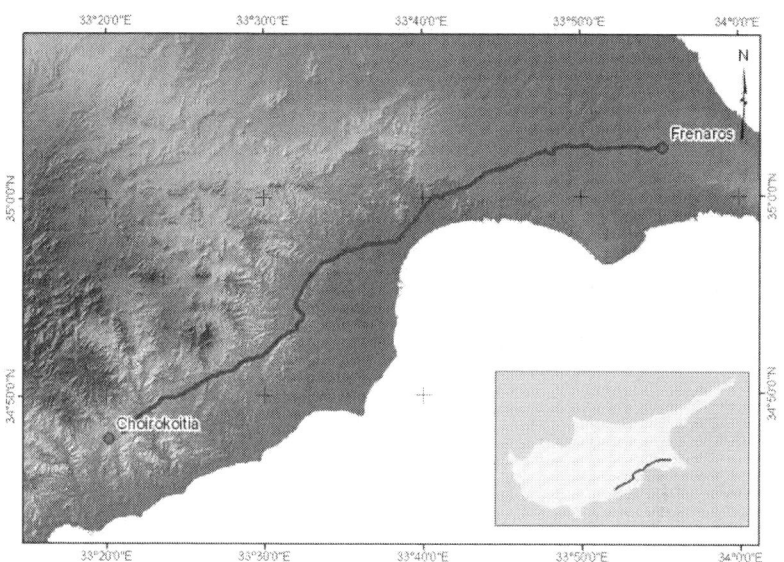

Figure 3. The *"Frenaros - Choirokoitia"* water pipe (solid line) used as the case study area.

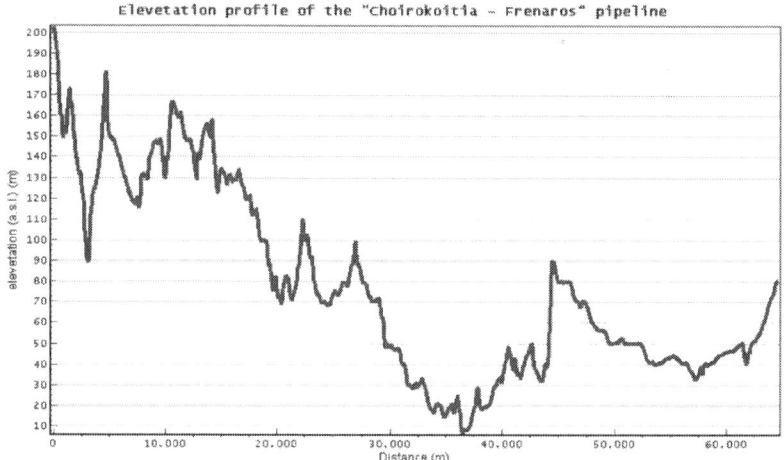

Figure 4. The elevation profile of the "Frenaros - Choirokoitia " waterpipe.

Gravity-Driven Water Flow in Networks Theory and Design

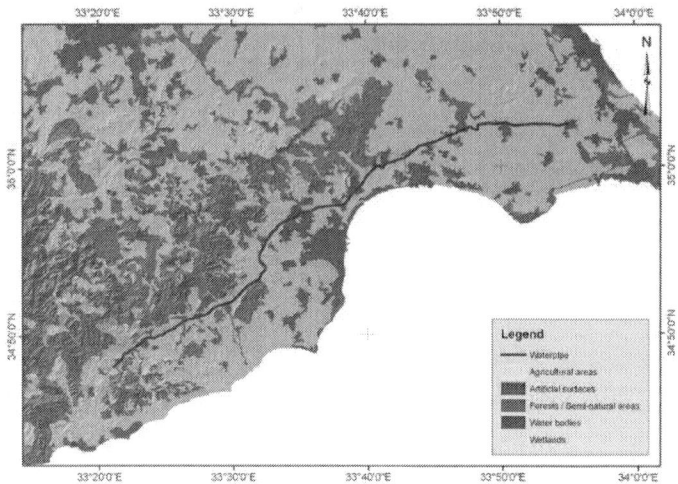

Figure 5. CORINE 2000 land use (Level 1) in the area of interest (*"Choirokoitia-Frenaros"* waterpipe)

During the period 2007 to 2010, three major leakages were observed along different sections of the pipe (Figure 6). The locations of these leakages were not detected until 2 months after the leakage occurred due to the difficulty of the local authorities in identifying the problematic areas. The leakages occurred during 2007; 2008 and 2010; further details for these events are presented in Table 1.

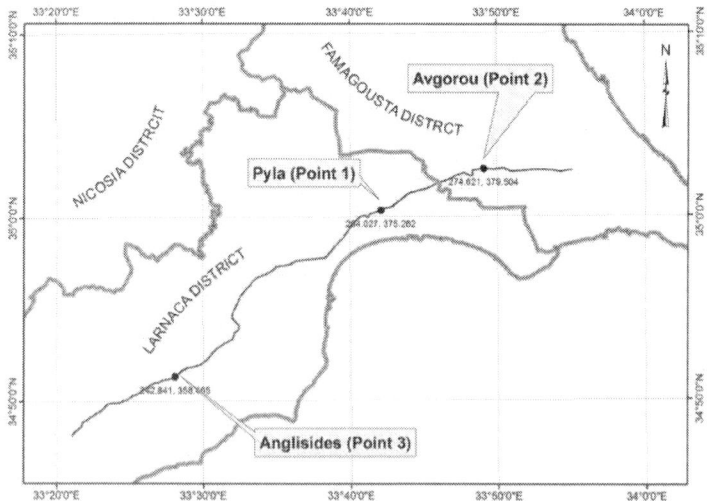

Figure 6. The "Frenaros - Choirokoitia" waterpipe (in blue). Points 1-3 indicate the three areas were water leakages have been reported.

Table 1. The leakages of the Frenaros – Choirokoitia water pipeline

Point	Position	Name	Date of pipe fixing
Point 1	Km 43.265*	Pyla Area	20-07-07
Point 2	Km 55.346*	Avgorou area	18-02-10
Point 3	Km 12.769*	Anglisides area	17-09-08

* Km positions along the pipeline, starting point Choirokoitia

METHODOLOGY

The detection of the footpirnt of the *"Southern Conveyor Project"* was made based on interpretation techniques. The interpretation was conducted using free data from Google Earth database and using high resolution satellite images. Several histogram enhancement techniques were applied along with filters in order to improve the interpretation. As well, Principal Component Analysis (PCA) and classification techniques were also conducted.

In order to explore the capabilities of remote sensing for the detection of water leakages, two different methodologies were followed. For the *"Lakatameia"* waterpipe pilot study, ground spectroradiometric measurements were taken using a handheld spectroradiometer. A leakage event was created by filling several sections of the pipeline with water so that ground spectral signatures could be taken before and after the leakage. Spectroradiometric data were also recorded from different heights using a low altitude system. In this way, spectral signatures were able to simulate variation in spatial resolution (pixel size) before any other further application.

For the *"Frenaros - Choirokoitia"* water pipe case study, three major leakages have been recorded (seeTable 1). Several Landsat 7 ETM+ medium resolution images, showing each leakage before and after the day the leakage was repaired, were used. A geometric and radiometric calibration of the images was performed, following by a multi-temporal analysis of all dataset based on either false composites or vegetation indices.

RESOURCES

In this section, the resources and processing used for each case study are presented. The resources are grouped into three main categories: (a) high resolution satellite data used for the *"Southern Conveyor Project"* area; (b) spectroradiometric ground data used for the *"Lakatameia"* pipeline and (c) medium resolution satellite data used for the *"Choirokoitia- Frenaros"* pipeline.

High Resolution Satellite Data

IKONOS high resolution satellite images were used for the detection of the buried water pipe. The IKONOS sensor, launched in 1999, was the first high-resolution satellite imagery with a spatial resolution of less than 4m. In addition, free RGB satellite images from the Google Earth database were explored and analyzed (23-10-2003; 13-06-2004; 29-05-2008; 30-05-2009) (Figure 7).

Figure 7. IKONOS satellite image used for the detection of the buried water pipe (left) and free Google Earth images of the area (right).

Spectroradiometric Data

Spectroradiometric hyperspectral measurements were carried out using the GER 1500 field spectroradiometer (Figure 8a). The GER 1500 spectroradiometer records electromagnetic radiation between 350 nm to 1050 nm (visible and near infrared part of the spectrum) A calibrated Spectralon panel, with ≈100% reflectance, was also used simultaneously to measure the incoming solar radiation. The spectralon panel measurement was used as a reference, while the measurement over the crops as a target. Therefore, reflectance for each measurement can be calculated using the following equation (1):

Reflectance = (Target Radiance / Panel Radiance) x Calibration of the panel   (1)

In order to avoid any errors due to changes in the prevailing atmospheric conditions (Milton et al. 2009), the measurements over the panel and the target were taken within minutes of each other. The coordinates of the measurements were mapped using a Global Navigation Satellite Systems (GNSS) (Figure 8b).

(a)   (b)

Figure 8. (a): The GER 1500 spectroradiometer used for the collection of ground measurements and (b): the GNSS used for mapping the pipeline

In addition, spectroradiometric measurements were taken from a low altitude system (Figure 9). The spectroradiometer was attached to the air balloon and raised over the pilot study area. Measurements were taken at several heights in the pilot study area and also in the surrounding area in order to compare their spectral signature profiles. As the airborne system was raised, the pixel size in the ground increased. Table 2 presents some characteristic heights where the pixel size corresponds to known satellite sensors.

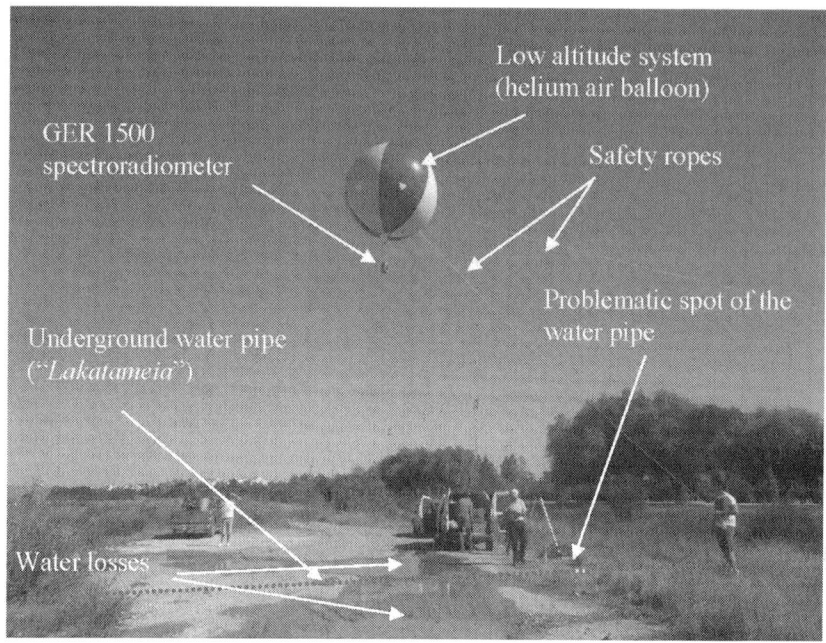

Figure 9. The low altitude system deployed over the leakage in the *"Lakatameia"* waterpipe.

Table 2. Pixel size from different heights using the low altitude system. The right column presents active satellite sensors with similar spatial resolution. Two lens with different field of view (FOV) have been be used in the GER 1500 spectroradiometer

Height from the ground	4° FOV (ground pixel - m)	8° FOV (ground pixel - m)	Satellite sensor
5	0.3	0.7	GeoEye (pan); WorldView-1
10	0.7	1.4	IKONOS (pan)
15	1	2.1	
20	1.4	2.8	ALOS (pan)
25	1.7	3.5	
50	3.5	7	IKONOS (multi)
75	5.2	10.5	ALOS (multi)
100	7	14	
150	10.5	21	Landsat (pan)
200	14	28	IKONOS (multi)

Hyperspectral measurements recorded from the GER 1500 instrument needed to be recalculated according to the characteristics of a specific multispectral satellite sensor. The authors modified these data to mimic Landsat 7 ETM+ satellite imagery based on Relative Spectral Response (RSR) filters since such data are freely distributed from the USGS. This data were used for the second case study (*"Frenaros - Choirokoitia"* waterpipe). RSR filters describe the instrument relative sensitivity to radiance in various parts of the electromagnetic spectrum (Wu et al. 2010). These spectral responses have a value of 0 to 1 and have no units since they are relative to the peak response (Figure 10). Bandpass filters are used in the same way in spectroradiometers in order to transmit a certain wavelength band and block others. The reflectance from the spectroradiometer was calculated based on the wavelength of each sensor and the RSR filter as follows:

$$R_{band} = \Sigma(R_i * R_SR_i)/\Sigma R_SR_i \qquad (2)$$

Where:

R_{band} = reflectance at a range of wavelength (e.g. Band 1)
R_i = reflectance at a specific wavelength (e.g R 450 nm)
R_SR_i = Relative Response value at the specific wavelength

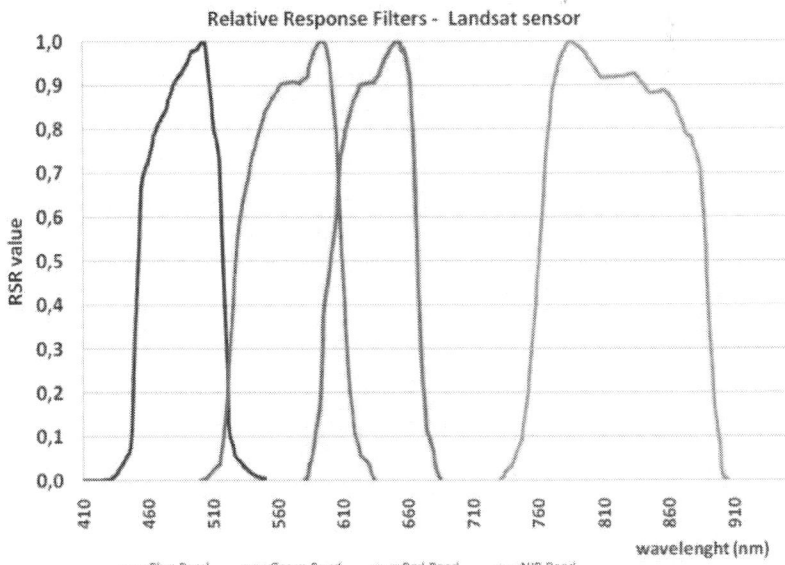

Figure 10. Relative Response filters for Bands 1-4 of Landat TM sensor (Alexakis *et al.* 2012)

Medium Resolution Satellite Data

Twelve medium resolution Landsat 7 ETM+ satellite images were used, dated before and after the local authorities fixed the leaks on the *"Frenaros-Choirokoitia"* pipeline (Figure 11; Table 3). ERDAS Imagine v. 10 software was used for the pre- and post-processing of satellite imagery. Pre-processing included geometric and atmospheric correction correction of the satellite imagery. Geometric correction of the satellite images was conducted using ground control points (GCPs), which included environmental features and ground coordinates. The Darkest Pixel (DP) atmospheric correction method was used, which is the most widely applied method of atmospheric correction that provides reasonable correction (Hadjimitsis et al., 2004; Hadjimitsis et al., 2009).

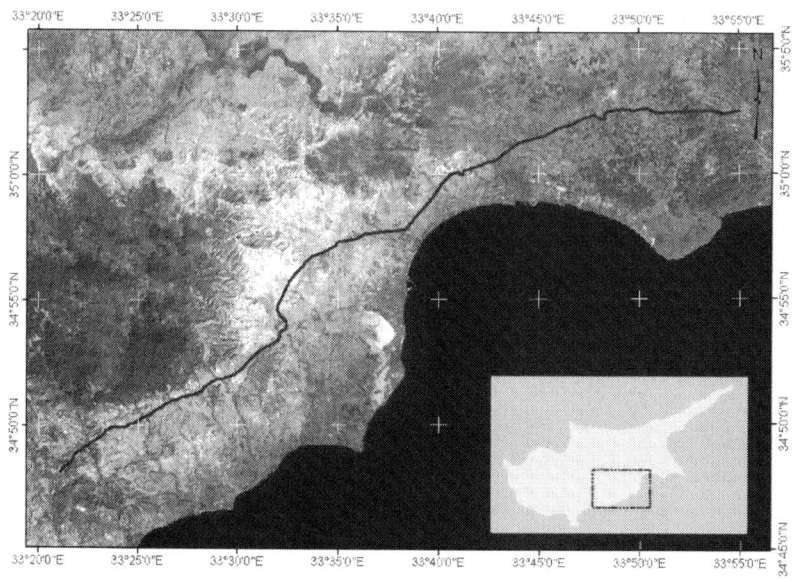

Figure 11. Landsat 7 ETM+ satellite image (28/07/2008) over the *"Choirokoitia-Frenaros"* water pipe.

After the necessary pre-processing steps, several vegetation indices were evaluated. False colour composites were also applied in order to detect the water leakages from the entire dataset. The evaluation was made not only in the three areas of interest (leakage problem) but also along the entire length of the water pipe. The results were mapped and statistical analysis was performed.

Table 3. Satellite images used for this study

no	Satellite	Overpass	no	Satellite	Overpass
1	Landsat ETM+	07-05-07	7	Landsat ETM+	14-09-08
2	Landsat ETM+	23-05-07	8	Landsat ETM+	30-09-08
3	Landsat ETM+	27-08-07	9	Landsat ETM+	16-10-08
4	Landsat ETM+	28-07-08	10	Landsat ETM+	22-12-09
5	Landsat ETM+	13-08-08	11	Landsat ETM+	07-01-10
6	Landsat ETM+	29-08-08	12	Landsat ETM+	13-04-10

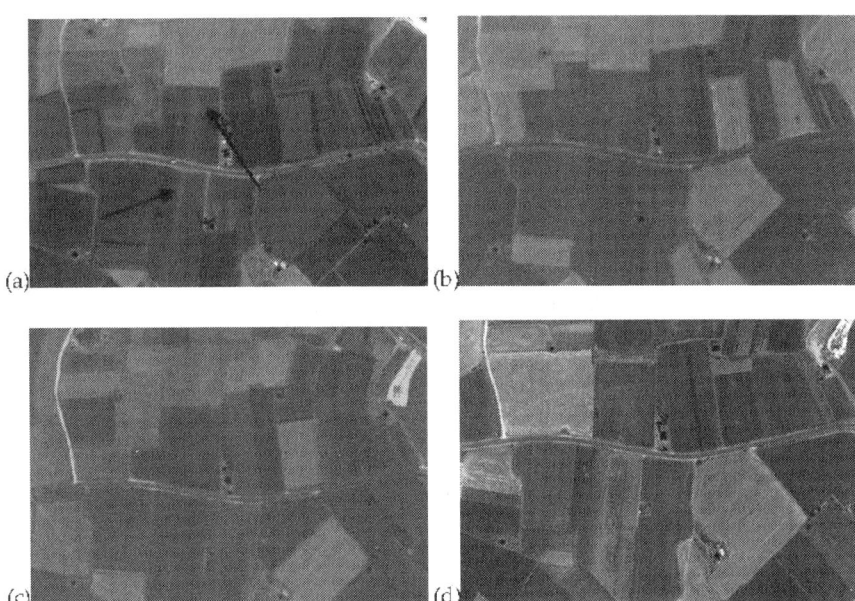

Figure 12. Google Earth satellite image used for the detection of the buried water pipe during different periods: (a): 23-10-2003; (b): 13-06-2004; (c): 29-05-2008; and (d): 30-05-2009.

RESULTS

"Southern Conveyor Project" Pipeline

The detection of the buried water pipe was initially performed using the multi-temporal Google Earth images (Figure 12). As shown, the success rate for the detection of the water pipe can vary depending on the period of observation. The interpretation could be performed much easier in areas with no coverage (bare soil) while in cultivated areas the interpretation was a difficult task. In addition, images taken just after rainfall or after watering crops, tend to provide better results since soil marks could be easily spotted.

Moreover, the tree pattern could reveal the footprint of the water pipe (see Figure 13). This pattern can be used for the detection of buried water pipes or can be used for monitoring possible problems resulting from tree roots.

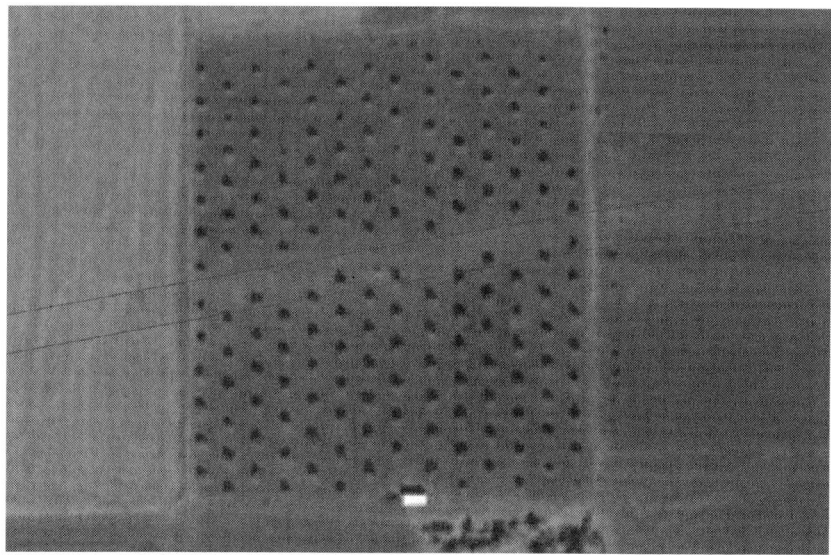

Figure 13. The footprint of the water pipe as a result of the tree pattern.

The IKONOS image used for this case study was able to maximize the visible footprint of the water pipe. Indeed, using the VNIR part of the spectrum and false colour composites (Figure 14) made possible the detection of both soil and crop marks. The IKONOS multispectral image was able to detect other parts of the water pipe network of the area, as shown in Figure 14 (right arrow). Spatial filter and PCA analysis applied to the image data (Figure 15) were able further to enhance the interpretation.

Figure 14. IKONOS VNIR-R-G pseudo colour composite

Figure 15. IKONOS 3 x 3 high pass filter (left) and PCA analysis (right)

In an attempt to evaluate if an automatic detection of such crop marks could be performed (e.g. classification), spectral profiles were examined. Spectral signatures from the image were evaluated as shown in Figure 15, which features areas of crop marks and of healthy vegetation. Scatter plots from these two areas (Figure 16) indicate that a spectral difference exists between these areas, especially in the VNIR part of the spectrum.

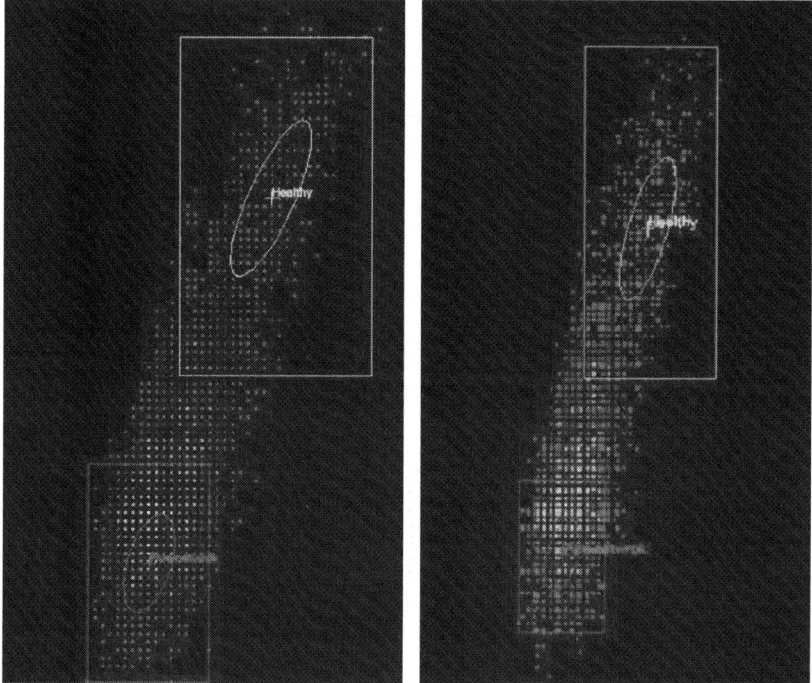

Figure 16. Scatter plots from crop marks (red square) and healthy vegetation (yellow square) for Bands 1-3 and Bands 1-4 combinations (left and right respectively).

"Lakatameia" Pipeline

The results found that water leakages could be monitored using remote sensing techniques. As shown in Figure 17, the spectral signatures of dry and wet soil is easily recognized in the visible range of the spectrum (400 - 700 nm) and in the very near infrared range (750-900nm). Wet soil tends to give 20-25% lower reflectance values compare to the dry soil. This difference is also maximized in the very near infrared range of the spectrum. Similarly, Figure 18 indicates spectral signature profiles of several targets before (dry) and after (wet) a leakage event. Similar findings also applied to vegetation. Dry grass tends to give approximately 5% reflectance in the green part of the spectrum (520-600nm) and 25% in the very near infrared (750-900nm) in contrast to 12% and 35% respectively for the wet grass.

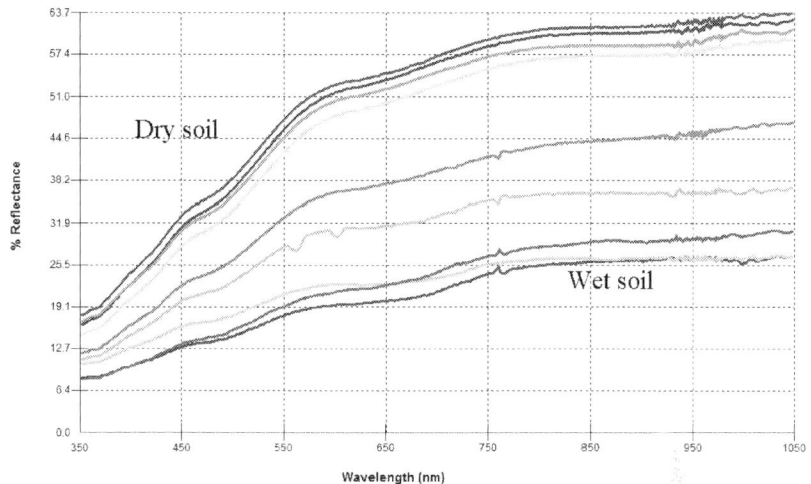

Figure 17. Ground spectral signatures over dry and wet soil in the *'Lakatameia'* pipeline

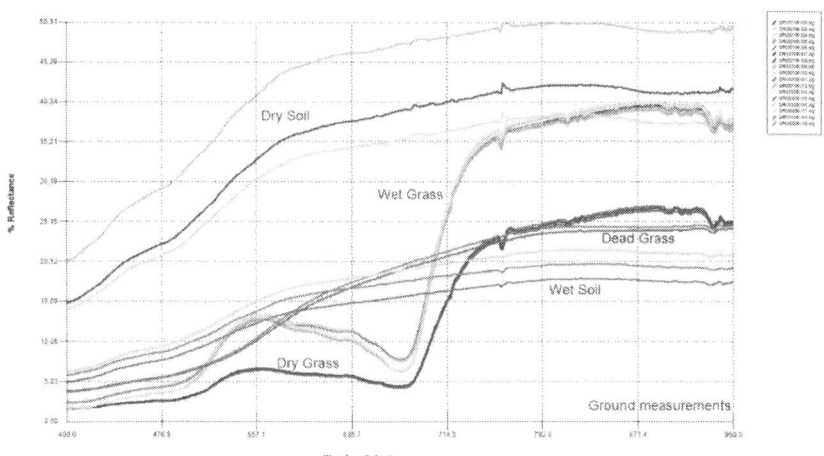

Figure 18. Ground spectral signatures of different targets in the *'Lakatameia'* pipeline

Figures 19 and 20 present the spectral signatures over the same areas from different heights, using the low altitude system. Reflectance initially increases as the system is raised above ground level (until 10 meters) while a small decrease of the reflectance is observed afterwards (16 meters) which can be associated with the larger area covered from the spectroradiometer. However it should be noted that these differences

(~5%) are similar to the total relative uncertainties of calibration for satellite sensors (within 5%) (Trishchenko et al. 2002).

The above results are well supported in the literature. Nocita et al. (2011), Ouillon et al. (2002), Dobos (2003), Kaleita et al. (2005) and Garcia-Rodriguez (2011) found that moisture affects the reflectance value of soil. There is a notable decrease in reflectance with increasing moisture in the ground (Bowers and Hanks, 1965; Baumgardner et al., 1985; Twomey et al., 1986; Ishida et al., 1991; Whiting et al., 2000; Bogrekci and Lee, 2005; Lesaignoux et al. 2007). However, the rate of decrease in relative reflectance becomes more moderate with increasing ground moisture, since at very high moisture contents, the soil is already quite dark and further moisture added to the soil has less of an effect on the reflectance (Kaleita et al., 2005). Moisture dominates the spectral reflectance of soils in the 340-2500 nm wavelengths (Somers et al., 2010; Bogrekci and Lee, 2005). Moisture affects the reflection of shortwave radiation from ground surfaces in the visible and near-infrared - VNIR (400-1100nm) and shortwave infrared - SWIR (1100-2500nm) regions of the spectrum (Bowers and Hanks, 1965; Skidmore et al., 1975). It is notable that, although precipitation affects the reflectance value for each target, it does not change the typical spectral signature between wet and dry conditions (Philpot, 2010).

The results indicate that the detection of a leakage event is possible using remote sensing techniques. Indeed, the use of the very near infrared range of the spectrum can be used on areas with bare soil or with vegetation. The findings from this pipeline were therefore compared with data from actual cases studies of water leakage in the *'Freanaros-Choirokoitia'* pipeline.

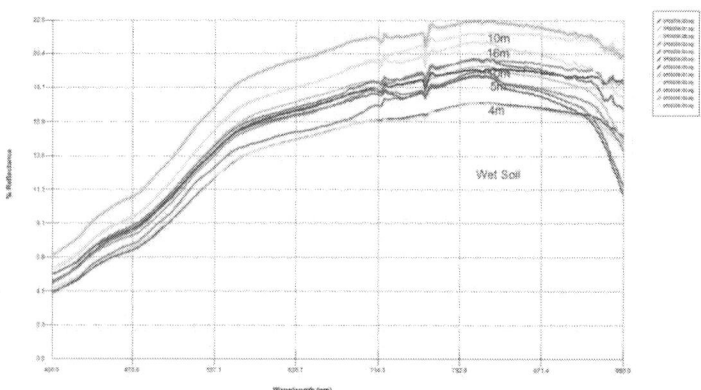

Figure 19. Spectral signatures of wet soil in the *'Lakatameia'* pipeline at different heights using the low altitude system

RESULTS

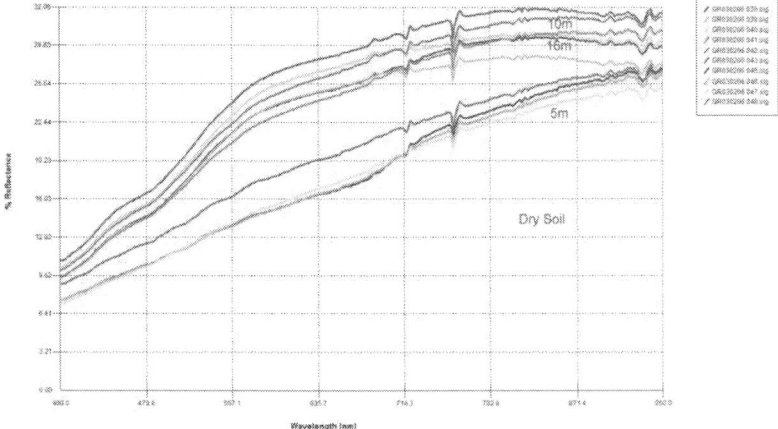

Figure 20. Spectral signatures of dry soil in the *'Lakatameia'* pipeline at different heights using the low altitude system

"Frenaros — Choirokoitia " Water Pipe

Based on the findings of the *"Lakatameia"* water pipe, satellite images where used for the detection of known water leakages using archive satellite images. In order to examine the capabilities of satellite remote sensing images for the detection of water leakages, several algorithms and analyses were carried out. At first, reflectance values of all datasets (see Table 3) were calculated based on the metadata file using equations 3 and 4. Following this, several vegetation indices were calculated. In addition, different false colour composites were produced to assess the ability of the system to detect the known leakages from the satellite images.

For Point 1 at *Pyla* area, leakage detection was difficult using medium resolution images. Monitoring of the pipeline using the red and the near infrared part of the spectrum for Point 1 did not reveal any significant changes of reflectance due to the water leakage. Similarly, vegetation indices (NDVI) did not show any differences for Point 3 (*Anglisides* area). However, for Point 2, Landsat 7 ETM+, promising results were found. As shown in Figure 21, the Landsat satellite image dated January 7, 2010, tends to give higher vegetation index values, prior to the water leakage being repaired on February 18, 2010. However, the above hypothesis is applicable to other areas of the water pipe as well. The above results have shown the limitations of using medium resolution satellite images for the detection of water leakages, especially when these are rare and small.

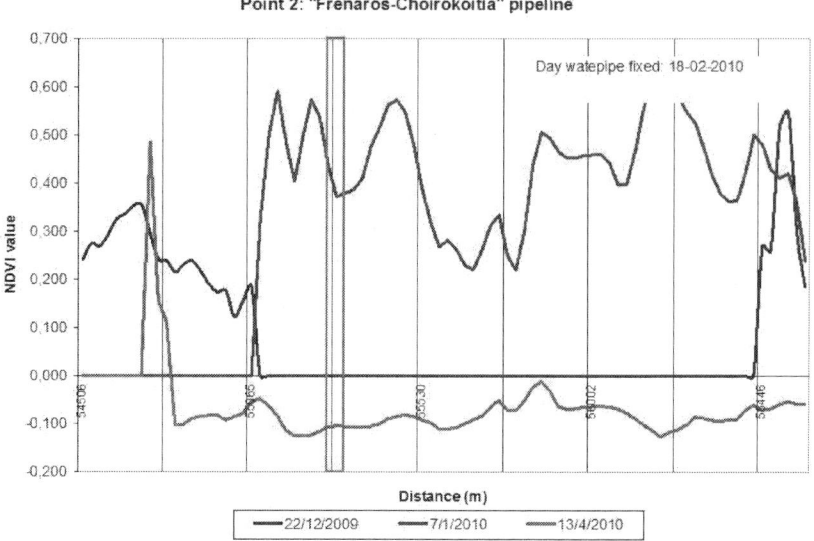

Figure 21. NDVI values using Landsat 7 ETM+ images used over Point 2. The red square highlights the area where the leakage was observed.

In an effort to explore further the information extracted using satellite data the three pilot areas were examined separately. Three vegetation indices, the

Normalized Difference Vegetation Index (NDVI);
Soil Adjusted. Vegetation Index (SAVI) and the
Ratio Vegetation Index (RVI) were calculated based on the formulas shown in equations 5, 6 and 7.

$$(pNIR- pred) / (pNIR+ pred)$$

(3)

$$(1+0.5) (pNIR- prb)/(pNIR+ pred+0.5)$$

(4)

$$pred/pNIR \tag{5}$$

Where:

p_{NIR} is the near infrared reflectance
p_{red} is the red reflectance

Figure 22 presents the NDVI development during the examined 12 dates of satellite overpasses (seeTable 3). Figure 23 presents the SAVI development during the examined 12 dates of satellite overpasses.

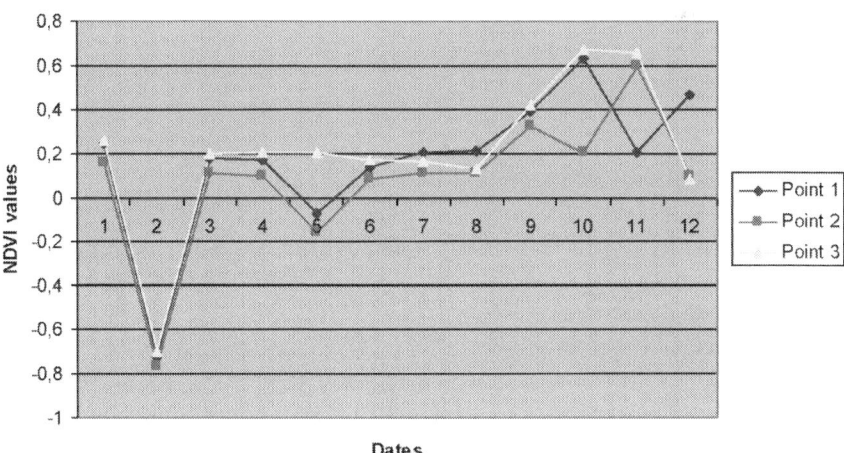

Figure 22. NDVI refl. (calculated with Reflectance values) development during the examined 12 dates (Landsat images) in Points 1, 2 and 3

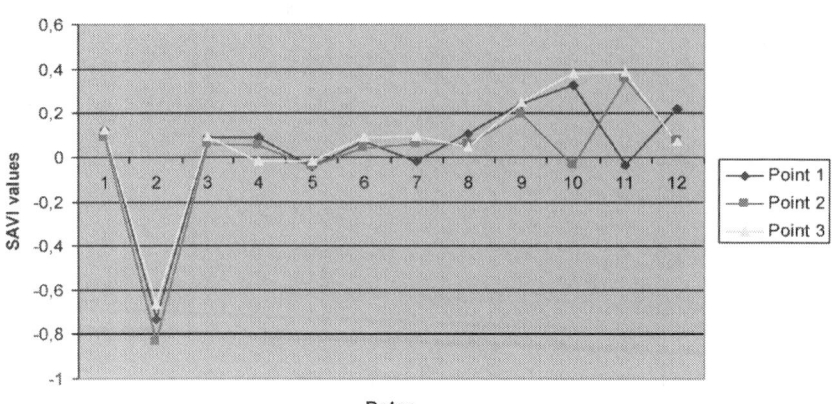

Figure 23. SAVI refl. (calculated with Reflectance values) development during the examined 12 dates (Landsat images) in Points 1, 2 and 3

Based on the graphs of Figure 22, the NDVI values present the following pattern: during May 2007, in all 3 points of the known water leakage, NDVI decreases significantly with values close to -0.8, when almost in all cases NDVI is above zero with similar values. After September 2008, the NDVI values increase until April 2010 when they decline again. Such results indicate that the vegetation of the area around the study points reflects soil moisture resulting from rainfall as it can differentiate according to season. Detailed examination of each point related to the pipeline repair indicates that NDVI in Points 1 and 2, water leakage ceased just after the 2nd and the 11th date in correspondence: (a) Point 1: -0,72 and 0,17 for days 2 and 3 and (b) Point 2: 0,60 and 0,09 for days 11 and 12 respectively.

For Point 1, there is a significant change of NDVI value before and after the repair date of the pipeline. In Point 2, the NDVI value decreased significantly (from 0, 60 to 0, 09) following the repair of the pipeline.

However, in Point 3 there is no significant change of the NDVI value before and after the repair date of the pipeline. Although there is a slight decrease in NDVI values immediately following the repair, there is a significant increase within 2 weeks: Point 3: 0,16; 0,13 and 0,42 for days 7 -9 respectively.

The results indicate that only at Point 2 is there a significant decline of NDVI values as a result of lack of soil moisture around the pipe. Another factor can be that due to the temporal difference between the two measurements, of 7 January 2010 and 13 April 2010, respectively, as lack of rainfall may have resulted in moisture evaporation. The same conclusion is reached with SAVI data (Figure 23). The value of SAVI in Point 2 was 0,35 in January 2010 and declined to 0,07 just after the pipeline repair.

Figure 24 presents RVI data which were calculated using equation 7. The RVI index indicates the effect of soil moisture around Point 2. The RVI value in Point 2, in January 2010 was 4,02 and after the pipeline repair, it decreased to 1,21. It seems that the vegetation developed on the soil around Point 2, and subsequently dried after the repair of the water pipe and the evaporation of the soil water.

In addition, meteorological data provided from the Meteorological Service of Cyprus, indicate that significant rainfall was recorded on 25, 26 and 27 February 2010, after the pipe line repair date of Point 2 (18 February, 2010). During March and April of 2010, only 1.0 and 2.1 mm of rain were recorded for the same location. Such information provides additional validation that the main factor affecting the NDVI, SAVI and RVI values is the presence or absence of vegetation as a result of soil moisture before and after the pipeline leakage repair. Regarding Point 3, in the Anglisides area, September precipitation data did not affect the pipe leakage since no significant rainfall was recorded before and after the pipeline repair (17 September, 2008).

Points 1,2,3: "Frenaros-Choirokoitia" pipeline

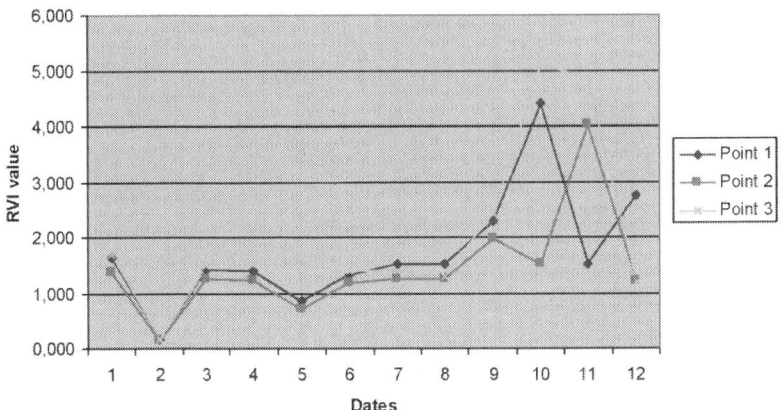

Figure 24. RVI refl. (calculated with Reflectance values) development during the examined 12 dates (Landsat images) in Points 1, 2 and 3

DISCUSSION AND REMARKS

Remote sensing techniques have been found to be effective both for the detection of the water pipes and for the detection of water leakages. The preliminary results of this study have shown that remote sensing techniques are able to detect areas of the pipeline with water leakages. Ground spectroradiometric data along with the low altitude spectroradiometer system indicate significant differences in the reflectance values in areas where leakage is observed. In addition, crop and soil marks can be used for mapping the actual footprint of the water pipe.

Although the use of medium resolution satellite images for monitoring extensive pipelines may be problematic, such as in Points 1 and 3 in the "*Franaros - Choirokoitia*" pipeline, this may be due to the spatial resolution of the specific satellite images. However, promising results have been also reported (i.e. Point 2 in the "*Franaros - Choirokoitia*" pipeline), where a major leakage was observed.

In addition, remote sensing techniques can be used on a systematic basis to monitor specific problematic areas of a water network by using time-series satellite images. Future research will investigate additional ground based geophysical methods to provide a competent system for monitoring existing water pipe networks, such as electrical resistance tomography and ground penetrating radar. The resulting data can be integrated into a Geographical Information System which can be used by local authorities.

ACKNOWLEDGEMENTS

The results reported here are based on findings of the Cyprus Research Promotion Foundation project "ΑΕΙΦΟΡΙΑ/ΦΥΣΗ/0311(ΒΙΕ)/21": Integrated use of space, geophysical and hyperspectral technologies intended for monitoring water leakages in water supply networks in Cyprus. The project is funded by the Republic of Cyprus and the European Regional Development Funds. Thanks are also given to the Remote Sensing and Geo-Environment Laboratory of the Department of Civil Engineering & Geomatics at the Cyprus University of Technology for its continuous support (http://www.cut.ac.cy).

REFERENCES

1. A Agapiou, D. G Hadjimitsis, D. D Alexakis, 2012aPerformance of vegetation indices for supporting ground and satellite remote sensing archaeological investigations. Remote Sensing. 4doi:10.3390/rs40x000x
2. A Agapiou, D. G Hadjimitsis, D Alexakis, A Sarris, 2012bObservatory validation of Neolithic tells ("Magoules") in the Thessalian plain, central Greece, using hyperspectral spectro-radiometric data, Journal of Archaeological Science, 39514991512doi.org/10.1016/j.jas.2012.01.001.
3. A Agapiou, G. D Hadjimitsis, A Sarris, K Themistocleous, G Papadavid, 2010Hyperspectral ground truth data for the detection of buried architectural remains, Lecture Notes in Computer Science, 6436318331
4. M Ahadi, S. M Bakhtiar, 2010Leak detection in water-filled plastic pipes through the application of tuned wavelet transforms to Acoustic Emission signals. Applied Acoustics, 71634639
5. D Alexakis, A Agapiou, D. G Hadjimitsis, A Sarris, 2012Remote sensing applications in archaeology. Remote Sensing / Book 2 (979-9-53307-231-8Book edited by: Boris Escalante.
6. A Bannari, D Morin, A. R Huette, F Bonn, 1995A review of vegetation indices. Remote Sensing Reviews, 1395120
7. M. F Baumgardner, L. F Silva, L. L Biehl, E. R Stoner, 1985Reflectance properties of soils. Advanced Agronomy, 38144
8. I Bogrekci, W. S Lee, 2006Effects of Soil Moisture Content on Absorbance Spectra of Sandy Soils in Sensing Phosphorus Concentrations Using UV-VIS-NIR Spectroscopy. Transactions of the American Society of Agricultural Engineers, 4911751180
9. S. A Bowers, R. J Hanks, 1965Reflection of radiant energy from soils. Soil Science, 2130138
10. L. S Burn, P Davis, D Desilva, B Marksjo, S. N Tucker, C. J Geehman, 2001The Role of Planning Models in Pipeline Rehabilitation. Plastic Pipes XI,36th September 2001. Munich, Germany.
11. E Dobos, 2003Albedo, Encyclopedia of Soil Science, 13DOI:E-ESS 120014334.
12. S Eyuboglu, H Mahdi, dan Al-Shukri, H. (2003Detection of Water Leaks using Ground Penetrating Radar. The 3rd International Conference on Applied Geophysics. December 812Orlando, Florida: Environmental and Engineering Geophysical Society.
13. S. N Faidrullah, 2007Normalized Different Vegetation Index for Water Pipeline Leakage Detection. In: The 28th Asian Conference on Remote Sensing 20071216November 2007, PWTC, Kuala Lumpur, Malaysia.

14. García RodríguezJ. N. (2011Changes in Spectral Slope due to the Effect of Grain Size and Moisture in Beach Sand of Western Puerto Rico. Accessed 21 October, 2011: http://gers.uprm.edu/pdfs/topico_johanna2.pdf.
15. D. G Hadjimitsis, C. R. I Clayton, A Retalis, 2009The use of selected pseudo-invariant targets for the application of atmospheric correction in multi-temporal studies using satellite remotely sensed imagery, International Journal of Applied Earth Observation and Geoinformation, 11192200DOI:j.jag.2009.01.00.
16. D. G Hadjimitsis, C. R. I Clayton, V. S Hope, 2004An assessment of the effectiveness of atmospheric correction algorithms through the remote sensing of some reservoirs. International Journal of Remote Sensing, 2536513674DOI:
17. D. G Hadjimitsis, K Themistocleous, C Achilleos, 2009Integrated use of GIS, GPS and Sensor Technology for managing water losses in the water distribution network of the Paphos Municipality in Cyprus, STATGIS 2009, Milos-Greece, 1719June, 2009.
18. M Heathcote, D Nicholas, 1998Life assessment of large cast iron watermains. Urban Water Research Association of Australia (UWRAA), Research Report 146
19. Y Huang, G Fipps, J. S Maas, S. R Fletcher, 2010Airborne remote sensing for detection of irrigation canal leakage. Irrigation and Drainage, 59524553
20. O Hunaidi, P Giamou, 1998Ground Penetrating Radar for detection of leaks in buried plastic water distribution pipes. Seventh International Conference on Ground Penetrating Radar,Lawrence, Kansas, USA, 2730
21. T Ishida, H Ando, M Fukuhra, 1991Estimation of complex refractive index of soil particles and its dependence on soil chemical properties. Remote Sensing of Environment, 38173182
22. A. L Kaleita, L. F Tian, M. C Hirschi, 2005Relationship between soil moisture content and soil surface reflectance. Transactions of the American Society of Agricultural Engineers, 4819791986
23. A Lesaignoux, S Fabre, X Briottet, A Olioso, E Belin, T Cedex, 2009Influence of surface soil moisture on spectral reflectance of bare soil in the 0.415m domain. Proceedings of the 6th EARSeL SIG IS workshop imaging spectroscopy.
24. E. J Milton, M. E Schaepman, K Anderson, M Kneubühler, N Fox, 2009Progress in Field Spectroscopy. Remote Sensing of Environment, 11392109
25. M Nocita, A Stevens, B Van Wesenmael, 2011Improving spectral techniques to determine soil organic carbon by accounting for soil moisture effects. The Second Global Workshop on Proximal Soil Sensing, Montreal.

REFERENCES

26. S Ouillon, Y Lucas, J Gaggelli, 2002Hyperspectral detection of sand. Presentation at the Seventh International Conference on remote sensing and coastal environments. Miami Florida, May 2022
27. W Philpot, 2010Spectral Reflectance of Wetted Soils. Proceedings of ASD and IEEE GRS; Art, Science and Applications of Reflectance Spectroscopy Symposium, Vol. II.
28. J. M Pickerill, T. J Malthus, 1998Leak detection from rural aqueducts using airborne remote sensing techniques. International Journal of Remote Sensing, 1924272433
29. Z Poulakis, D Valougeorgis, C Papadimitriou, 2003Leakage detection in water pipe networks using a Bayesian probabilistic framework. Probabilistic Engineering Mechanics, 18315327
30. A Sarris, N Papadopoulos, A Agapiou, C. M Salvi, D. G Hadjimitsis, A. W Parkinson, W. R Yerkes, A Gyucha, R. P Duffy, 2013Integration of geophysical surveys, ground hyperspectral measurements, aerial and satellite imagery for archaeological prospection of prehistoric sites: the case study of Vésztő-Mágor Tell, Hungary, Journal of Archaeological Science, 4014541470doi:j.jas.2012.11.001.
31. E. L Skidmore, J. D Dickerson, H Shimmelpfennig, 1975Evaluating surface-soil water content by measuring reflectance. Soil Science Society Annual Proceedings, 39238242
32. M. I Skolnik, 1990Radar Handbook, 2nd Ed. New York: McGraw-Hill.
33. B Somers, L Tits, W. W Verstraeten, P Coppin, 2010Soil reflectance modeling and hyperspectral mixture analysis: towards vegetation spectra minimizing the soil background contamination. Hyperspectral Image and Signal Processing: Evolution in Remote Sensing (WHISPERS), 14June 2010.
34. S. A Twomey, C. F Bohren, J. L Mergenthaler, 1986Reflectances and albedo diferences between wet and dry surfaces. Applied Optics, 25431437
35. P. A Trishchenko, J Cihlar, L Zhanqing, 2002Effects of spectral response function on surface reflectance and NDVI measured with moderate resolution satellite sensors. Remote Sensing of Environment, 81118
36. L Weifeng, L Wencui, L Suoxiang, Z Jing, L Ruiping, C Qiuwen, Q Zhimin, Q Jiuhui, 2011Development of systems for detection, early warning, and control of pipeline leakage in drinking water distribution: A case study. Journal of Environmental Sciences, 2318161822
37. M. L Whiting, L Li, S. L Ustin, 2003Estimating surface soil moisture in simulated AVIRIS spectra. Twelfth Annual Airborne Earth Science and Application Workshop. Jet Propulsion Laboratory, California Institute of Technology, Pasadena, California, February 2528

38. X Wu, T. J Sullivan, K. A Heidinger, 2010Operational calibration of the advanced very high resolution radiometer (AVHRR) visible and near-infrared channels. Canadian Journal of Remote Sesning, 36602616

CITATION

Diofantos G. Hadjimitsis, Athos Agapiou, Kyriacos Themistocleous, Giorgos Toulios, Skevi Perdikou, Leonidas Toulios and Chris Clayton (2013). Detection of Water Pipes and Leakages in Rural Water Supply Networks Using Remote Sensing Techniques, Remote Sensing of Environment - Integrated Approaches, (Ed.), ISBN: 978-953-51-1152-8, InTech, DOI: 10.5772/39309.

CHAPTER 3

An Analysis of the Interface between Evolutionary Algorithm Operators and Problem Features for Water Resources Problems. A Case Study in Water Distribution Network Design ☆

K. McClymont, E. Keedwell, , D. Savic

College of Engineering, Mathematics and Physical Sciences, University of Exeter, Harrison Building, North Park Road, Exeter EX4 4QF, United Kingdom

HIGHLIGHTS

- Explores the relationship between search operators and problem spaces in water resources problems.
- Standard mutation robust to most changes within a water distribution network.
- Problem specific operators less robust to looping in networks than mutation.
- Combinations of operators with problem specific operators improve performance markedly.

ABSTRACT

Evolutionary Algorithms (EAs) have been widely employed to solve water resources problems for nearly two decades with much success. However, recent research in hyperheuristics has raised the possibility of developing optimisers that adapt to the characteristics of the problem being solved. In order to select appropriate operators for such optimisers it is necessary to first understand the interaction between operator and problem. This paper explores the concept of EA operator behaviour in real world applications through the empirical study of performance using water distribution networks (WDN) as a case study. Artificial networks are created to embody specific WDN features which are then used to evaluate the impact of network features on operator performance. The method extracts key attributes of the problem which are encapsulated in the natural features of a WDN, such as topologies and assets, on which different EA operators can be tested. The method is demonstrated using small exemplar networks designed specifically so that they isolate individual features. A set of operators are tested on these artificial networks and their behaviour characterised. This process provides a systematic and quantitative approach to establishing detailed information about an algorithm's suitability to optimise certain types of problem. The experiment is then repeated on real-world inspired networks and the results are shown to fit with the expected results.

INTRODUCTION

Evolutionary algorithms (EAs) have been applied to a countless number of problems across a wide variety of disciplines. Their relative simplicity and ability to work well on new problems have led to them being adopted in fields as diverse as engineering, economics and robotics. As one would expect, EAs have also been applied to water resources problems with a large degree of success, for example in the fields of groundwater remediation (Piscopo et al., 2014), controlling channel bed morphology (Nicklow et al., 2003), determining the hydraulic characteristics of production wells (Jha et al., 2004) and in particular to the field of water distribution network optimisation (e.g. Savic and Walters, 1997, Bi et al., 2014). A key aspect of EAs is that they have a number of parameters to set when first considering a new problem (e.g. population sizes, mutation and crossover rates, selection pressure etc.) and this usually means that a period of parameter tuning is necessary to deliver acceptable performance. To alleviate this research has been carried out on algorithms to adapt these parameter settings automatically, removing the need for the parameter tuning period. A natural extension to this process then is to consider whether not just the parameters, but the operators themselves might be selected in an adaptive fashion, leading to the field known as selective hyperheuristics (Burke et al., 2013). In essence, this field explores the potential for algorithms to function beyond the strict application of selection, mutation and crossover phases and aims to develop a more dynamic approach with a greater number of operators to produce better results with less human intervention. However, to develop a suitable pool of operators from which a set might be chosen we must first understand the relationship between problem characteristics and operator function. The work herein introduces a process for exploring this interaction, and provides empirical results on a range of different artificial and real-world problems using water distribution networks as an example case study.

INTRODUCTION

Water distribution network design problem

Water distribution networks (WDNs) represent one of the most complex and key infrastructures in use today and are responsible for the transportation of clean drinking water from reservoirs and storage tanks to industrial and residential consumers. Failure of these networks to adequately supply the demand can cause significant problems in the day-to-day running of businesses and homes.

A standard WDN is comprised of pipes, nodes (junctions and demand points), hydraulic devices (such as pumps) and sources (tanks and reservoirs) that constitute the entire infrastructure that delivers water from the source (e.g., reservoir) to various locations where it is drawn from the network for consumption (e.g., residential housing or industrial sites). With increasing demand and tighter regulation, water companies continue to search for more optimal operations and improvements in their networks and so have in the last two decades looked towards emerging optimisation methods to help solve their problems. Real-world WDNs are complicated structures that require constant operational management, maintenance and rehabilitation. In order to satisfy consumer demand, the networks must be constructed with a good layout that connects to all points of demand and should provide the best possible hydraulic conditions and operational requirements all whilst minimising network cost. This is known as the WDN design problem.

The WDN design problem is known to be an NP combinatorial problem (Yates et al., 1984). Even for relatively small networks, the number of possible combinations of pipes is very large which makes enumeration of all the possible designs impossible. If, for example, there were six potential sizes for each pipe in a network of just thirty pipes, there would be $6^{30} = 2.21 \times 10^{23}$ possible combinations – far more than is possible to fully enumerate within a reasonable time. This basic complexity is further compounded when advanced controls such as pump scheduling and valve operations are considered in combination with the much larger models (e.g. 1000s of pipes) that are likely to be found in the real world. Finally, when the potential for independency in looped structures and the non-linearity of the hydraulic equations is included, it becomes clear the WDN design problem is difficult non-linear, multi-modal problem. It is because of this that researchers and practitioners look to more advanced meta-heuristics to optimise their WDN designs.

Optimisation of WDNs

Since the first application of optimisation methods (Blum and Roli, 2003) to the problem of water distribution network (WDN) design researchers have collectively established a large body of literature on the subject (Marchi et al., 2014). The majority of these studies are focused on the application of novel optimisation methods to this problem, novel formulations of the problem (McClymont et al., 2013), or case studies of real-world instances. In addition, these studies have often employed or proposed new meta-heuristic methods; predominantly those from Evolutionary Computation (EC) (Coello et al., 2007) and variants of Evolutionary Algorithms (EAs) (Laumanns et al., 2000). More recently, there has been a shift in focus to hybrid (Keedwell and Khu, 2005) or more adaptive methods (Afshar, 2006) for the optimisation of these problems such as multi-method search (Vrugt and Robinson, 2007, Vrugt et al., 2009 and Raad et al., 2010) and selective hyper-heuristics (McClymont et al., 2013).

The work presented here investigates search operators and their interaction with features in the fitness landscape for water distribution network optimisation, the first time this has been attempted, although attempts have been made to undertake studies of a similar nature in other domains (Franchini and Galeati, 1997). In a similar study, Zecchin et al. (2012) investigated Ant Colony Optimisation Algorithms in relation to Water Distribution Network problem characteristics and also highlighted the importance of these studies. Indeed, the optimisation of WDNs by Evolutionary Algorithms (EAs) in the early 1990s (Walters and Lohbeck, 1993, Simpson et al., 1994 and Savic and Walters, 1995) was the start of a wider effort to find new, more efficient and effective optimisation techniques for this difficult real-world problem.

Performance analysis

In the search for better optimisation methods, papers frequently attempt to analyse the performance of meta-heuristics by applying them to a set of large, realistic WDNs (Walters et al., 1999 and Cheung et al., 2003) or other water networks, such as in Fu et al. (2008). This experimental method provides vital information on the scalability of the proposed techniques. However, when considering the use of adaptive techniques to select operators it is important to understand the impact that individual search space features have on the behaviour and suitability of a method to that type of search space. What is required is to establish a fundamental understanding of the effect of different WDN features and landscape attributes on optimisation methodologies and lay the ground work for the new approaches described later. To make confident assertions about the true behaviour, such as the explorative or exploitative search of an algorithm, quantitative analysis of the algorithms is required (Deb and Jain, 2002 and Merz, 2004).

Furthermore, while these works are important in developing better techniques for solving this class of difficult and constantly evolving real-world problems (McClymont et al., 2013), there has been relatively little work conducted on the analysis of the fundamental rules of how these optimisation methods behave under different conditions in the context of the WDN problem. Consider, for example, the significant differences in the hydraulic properties of a looped network versus a dendritic network (Walters and Lohbeck, 1993) or, similarly, a gravity fed network versus a network of pumps with tank storage. These variations in structural and hydraulic properties will result in very different optimisation search space landscapes for operators to traverse. Therefore, while one optimisation method might perform well on certain types of network, it is equally likely that it will perform less well on others. It is important to understand this relationship, between optimiser and problem, in order to make well-grounded claims about any one method's suitability for solving the WDN problem, or certain variants of it. This understanding will also help to guide algorithm and operator selection for this class of problems.

Problem and operator linkage

It is clearly shown by the No-Free-Lunch theorem (Wolpert and Macready, 1997) that not all optimisers are well suited to solving all problems. Similarly, it can be said that not all operators in an optimiser are well suited to solving all problems. This statement can be generalised somewhat to say that not all optimisation operators behave in the same way and therefore are not suited to all problems. The question therefore is: to what extent is it possible to ascertain a profile detailing the behaviour of an optimiser or its operator(s) and to determine how this profile relates to specific problems and problem features?

Malan and Engelbrecht (2013) provide some insight into the concept of characterising generalised fitness landscapes and the early work by Kauffman (1989) suggests adaptation based on these variances in the landscape is feasible. Furthermore, studies such as Moraglio and Poli (2004) show how specific operators can have definite and identifiable attributes and behaviours (which has also been commented upon in the traditional evolutionary literature (Wright, 1932)).

This work addresses some of the current challenges facing EA research and specifically EAs for hydroinformatics as outlined in Maier et al. (2014) and is primarily concerned with exploring methods for the "development of knowledge of the underlying searching behaviour of different search methodologies". The work also touches on the "development of knowledge of the fundamental characteristics of the problem being optimised at the level at which optimisation algorithms operate" a concept that is inextricably linked with that above, as noted by Maier et al. in the section "Maier et al. (2014)". The behaviour of an optimiser (or part thereof) is always in response to the landscape and features of the problem being explored. One does not climb a downhill and roll up a cliff-face.

The work presented below constitutes a novel approach to quantitatively analysing and comparing different Evolutionary Algorithm operators on the WDN design problem. The method extracts key generic attributes of the WDN problem which are encapsulated in the natural features of the problem, such as network topologies and variable types, on which different EA operators can be tested. The method is demonstrated using small exemplar networks designed specifically so that they isolate individual features. A set of operators are tested on these artificial networks and their behaviour characterised. The method provides results that can be used to understand what, if any, linkages exist between the performance of an operator and certain features of a WDN, for example, the presence of pumps or the existence of looping topology. Finally, the operators are tested using real-world inspired networks for which the presence of each WDN feature is assessed. The test is a means of providing a level of confidence in the accuracy of the information learned about each operator.

METHOD

This section details the method used to compare and analyse the behaviour of EA operators and their relationship to problem features, described in general terms. The method described below is applied in this study to the WDN design problem.

A common approach to comparing methods
The majority of studies in the literature approach algorithm analysis and comparison in the same way, following from early studies and their proposed methodologies (Whitley et al., 1996 and Zitzler et al., 2000).

Usually, a set of benchmark problems are selected; either real-world examples or manufactured mathematical constructs and the algorithms to be compared are configured to operate effectively on the selected problems. This can be achieved by control parameter tuning on a subset of the problems or by using standard parameter values. The algorithms are then applied to (i.e., optimise) each problem and the results are collected. If the algorithm is a stochastic optimisation method, such as a metaheuristic, then it is applied multiple times (known as trial runs) to each problem in order to collect a set of results that provide statistical information about performance. The median result is usually used for comparison for sets of trial run results. This process is illustrated in Fig. 1.

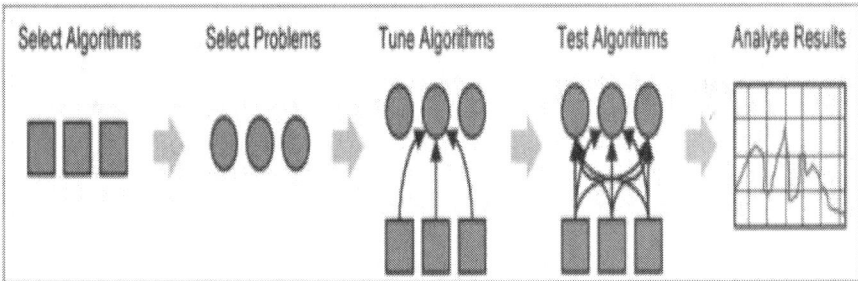

Figure 1. A common algorithm testing process. See Fig. 3 in experimental setup for the process used in this study.

The comparative run setup is common to most studies and provides a practical and easily repeatable means for collecting data about algorithm performance. Any measure can be used, from mean or best objective value to the distribution of solutions in parameter space. The algorithms can be compared using their final solutions alone or by comparing metrics over the whole optimisation process, such as their rate of convergence, if their optimisation scales (i.e., both generational) are compatible. The benefit of this setup is that the comparison is easily understood and reflects the actual optimisation process used when practically applied to real-world problems. The methodology presented in this paper takes this experimental approach and extends it to include preliminary steps to characterise the optimisers' performance in relation to specific features.

Method for characterising optimisers
Rather than applying the optimisers directly to problems and comparing overall performance, the method proposed here first attempts to characterise which problem features, if any, are related to an algorithm or

operator's performance to provide a better insight into the underlying causes of that performance.

Unlike the common approach shown in Fig. 1, this method follows a preliminary testing phase. The method is as follows: (1) select operators, (2) select problems, (3) identify problem features, (4) synthesize artificial problems, (5) test on artificial problems, (6) analyse results and determine linkages, (7) select the most appropriate operators for selected problems, (8) test on actual problems, (9) analyse results. This is conceptually shown in Fig. 2.

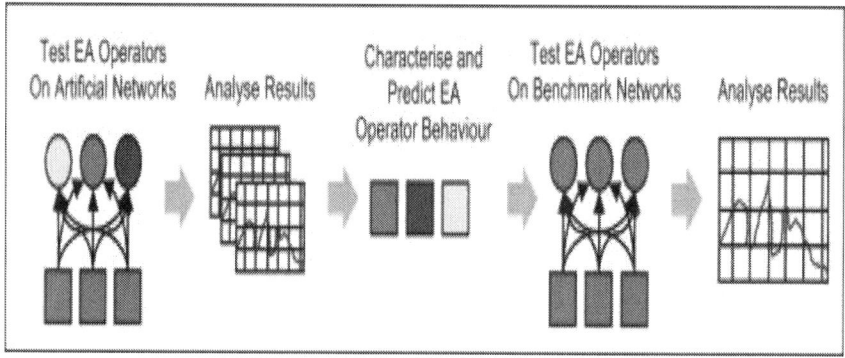

Figure 2. Illustration of the method for characterising EA operators given different problem features encapsulated in artificial water distribution networks.

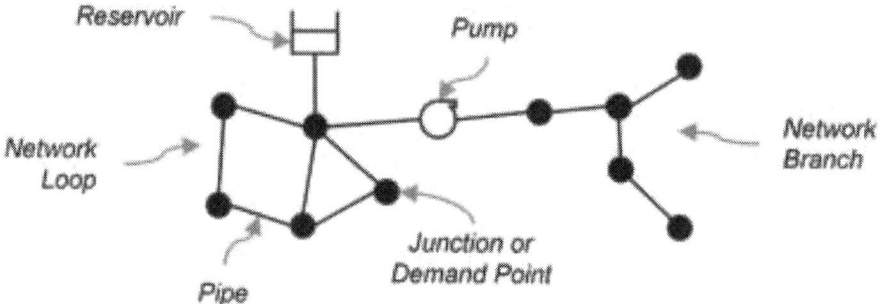

Figure 3. An example of a water distribution network schematic (topological illustration).

Selecting the operators and problems are common tasks and not described here. The process of identifying problem features is somewhat problem dependent. For example, the WDN problem examined in this study allows for easy categorisation of problem features as the different assets and topologies of the network are the driving force between the hydraulic differences and so objective function landscapes.

Defining problem features

It is common to use mathematical test problems in addition to using real-world benchmarks. These are well defined problems with known features and can be quickly evaluated to enable large scale studies. There are many test problems available in the literature, from those identified by Van Veldhuizen (Van Veldhuizen and Lamont, 1998 and Van Veldhuizen, 1999) through to more complex problem suites like the BBOB test functions (Hansen et al., 2010). This study is focused on the specifics of a real-world problem and so does not employ these test problems, although the notion of breaking down the problem into elements to construct new theoretical test problems is used here.

Furthermore, as is specified in Maier et al. (2014) (and outlined above), problem features are often described in terms of fitness landscapes: which can been seen to be the general properties of the functional mapping between parameter space and objective space. These features and the methods used to detect, analyse and capitalise on them are highly important in both the specific research of hydroinformatics but also to EA research and optimisation at large. This paper takes an alternative approach, exploring the specific features associated with the construction of a WDN. This is for two reasons: (1) the features are easily identifiable and understood by the general practitioner; and (2) the physical features of the WDN define the fitness landscape and so can be used interchangeably with those generalised features. Indeed, it should be noted that the categorisation of problems can be done on any set of features (including general ones) and is not limited to those used in this study.

Artificial problems

Synthesis of the artificial problems is similarly problem dependent. The aim is to take the set of identified features and synthesise artificial networks that represent each of these features independently and in combination with other features. For example, the presence of pumps and valves in a network present two key features in the WDN problem. Using these two features, four artificial networks can be created: no pumps or valves; valves but no pumps; pumps but no valves; both pumps and valves. These problems provide the basis for a systematic analysis of the affinity of any method to specific problem features. I.e., an EA may be well suited to problems with pumps, while others may perform better without pumps.

By separating the features and selectively recombining them, it becomes possible to identify correlations in performance.

Characterising & selection of optimisers

Using the set of artificial problems, each optimiser is applied to each problem (over a number of trial runs for a fairer comparison) to produce a matrix of results. The matrix of optimisers against problems (and by proxy, features) provides a means for detecting patterns between optimisers and problem features. Given the matrix of results, it becomes possible to make a more informed selection of optimisers to apply to the larger, more expensive real-world or benchmark problems. For expensive to evaluate problems, like the WDN problem, testing large numbers of optimisers on these problems is often unfeasible given the usually limited resources. The artificial problems, by contrast, should be smaller instances of the problem and faster to execute. Indeed, in this study, the artificial problems were orders of magnitude faster in computing time to evaluate which made the wider comparison of pairs possible.

THE WATER DISTRIBUTION NETWORK DESIGN PROBLEM

As stated above, in its basic form, a WDN is made up of pipes, junctions, demand points, hydraulic devices (such as pumps) and water sources (tanks and reservoirs). The network transports water from sources to various locations where it is drawn from the network for use. A topological illustration of an example network is given in Fig. 3.

Ideally, WDNs are constructed with a layout that connects to all demand points and provides the best possible pressure and water quality to satisfy demand. The design problem is primarily concerned with the sizing (diameters) of pipes in the network as well as scheduling of pumps and valve operations. Changing pipe sizes affects the hydraulic conditions in the network and the ability to serve the various demand points. Large pipe diameters are more expensive and so the aim of the WDN design problem is to reduce the cost of the network (minimize pipe sizes) while still satisfying customer demands (i.e., maintaining adequate pressure throughout the network). Similarly, turning pumps on and off alters the flow of the network and can increase the flow and pressure in areas of the network which are not well serviced by gravity feed alone, although the running of pumps has an additional operational cost in terms of energy consumption.

The single-objective WDN design problem is traditionally formulated as follows:

Minimise: $cost = \sum_{i=0}^{k}(c_i \times l_i) + \sum_{a=0}^{n}\sum_{b=0}^{m}(p_{a,b} \times e_b)$
Given: $head > 30m$, and $head < 40m$, and $velocity < 2.5ms^{-1}$

This formulation aims to minimise the combined cost of the pipe infrastructure and the energy costs associated with running each pump. The symbols are as follows: k = number of pipes in the network, ci = is the cost per metre of the selected diameter of pipe i (e.g. as shown in Table 1), li = the length of pipe in metres, n = the number of pumps in the network, m = the number of timesteps in a simulation, pa,b = a binary value determining whether pump a is on at timestep b, and eb = the cost of running a pump at timestep b in US dollars. Additional constraints on the feasibility of a design are given by head pressure and velocity requirements on the network. The head and velocity constraints are considered violated if, at any time point, the network exhibits values outside of the given constraints. As noted later, the network is solved using EpaNET and these values taken from the solver's results.

Table 1. Diameter and associated costs for pipes in the artificial networks.

Diameter (mm)	Cost ($/m)
150	25
200	75
250	125
300	175
350	225
400	275
450	325
500	375
550	425
600	475
650	500

WDN features

While much of the optimisation literature (beyond hydro-informatics) is concerned with the mathematical features of a problem, such as the mapping between parameters and objectives, and the landscape this creates, this study looks at the practical features present in the WDN problem. There are methods for testing for general landscape properties (multi-modality, deceptive minima, etc.) but this study looks at the base properties of the problem (topology and assets) that gives rise to these landscapes. The aim is to relate the operators to the problem, not general landscape properties. The WDN problem has a number of practical features, such as the topology of the network and the presence of different WDN assets, which directly relate to how easily the problem can be solved. The different topological features and assets are described below with a brief description of their impact on the problem.

Topology

There are two overarching or contrasting types of network topology: dendritic and looped networks. These extremes are described below, however it should be noted that most real-world networks are a combination of the two and are shown as 'hybrid' networks in the experiments below.

Dendritic networks have one or more water sources at the "centre" of the network. The network then extends from this source with pipes splitting into separate branches (like a tree) and serving different areas. The branches are not interconnected and so hydraulically separate to a degree. Each branch eventually terminates in an isolated end node or terminus. These networks are relatively simple to optimise as different areas have little effect on others.

Looped networks, in contrast, do not have end nodes and there are few branches in the network. Instead, the nodes are connected by "loops" that creates multiple routes of flow between the source and each demand node. These networks have more complex hydraulics and changing any one pipe can have a significant effect on many other pipes. The problem is therefore more complex with interdependencies between parameters. The networks are more robust than dendritic networks as demand nodes are not reliant on one route to the source and so can sustain a greater number of pipe failures and still function effectively. The types of loops can vary, with the most common being grid like structures that reflect the modern designs of cities. The placement of water sources also has an effect on the network type. A network could be supplied by a single source or a concentration of sources in the same area. Equally, the sources to a network can be distributed

THE WATER DISTRIBUTION NETWORK DESIGN PROBLEM

across the network. The latter being more common when tanks are introduced in to a network.

Assets

Modern WDN systems are built with a wide variety of different assets each of which will have an effect on the hydraulic conditions in the network, from the essential pipes, junctions and valves to more complex units such as strainers, valves, tanks, metres and detection units. This study explores the effect of core assets which are common to most urban WDNs, which are: reservoirs, tanks, flow control valves, and time controlled pumps.

Reservoirs and tanks provide the source water to a network and dictate much of the network structure and base flow conditions of the network. Gravity fed networks (where no pumps are required to maintain adequate pressure in the network) are entirely dependent on the placement and number of water sources in the network. Reservoirs represent the most common water source in the UK whereas tanks are often used (in combination with pumps) to introduce a greater level of resilience in a network which has pressure issues and or potential for water outages from the supplying reservoir(s). Networks with more than one source are often more robust to failure and able to provide more stable pressure conditions in the network.

Valves and pumps are devices designed to manage water flow and pressure in the network to improve conditions over those given by basic gravity fed systems. Pumps allow water to be pumped upstream to areas of the network which may not be serviced by gravity fed systems or have a reduced pressure. Pumps can also be timed to accommodate increased demand at certain periods in the daily cycle. Valves are used to prevent cyclical back-flow in looped systems and control water flow, being able to restrict flow at low demand periods and opening during higher demand periods. Both of these assets make the hydraulics of a network more complex and have an effect on the pipe diameters that are optimal for the network.

The artificial networks used in this study are built from a combination of secondary reservoirs, a tank, a flow control valve, and time controlled pumps. The diameters of the pipes and the pumping schedule are used as decision variables in the search.

EVOLUTIONARY ALGORITHM

An elitist Evolutionary Algorithm (EA) was used in this experiment to test different genetic operators. An EA is an iterative optimisation process which uses a population of candidate solutions which are varied to explore the search space. A standard EA takes a random initial population and then iterates through the three following processes: variation, evaluation, and selection (shown in Fig. 4). Variation is the process by which each of the existing candidate solutions are altered to create a new solution to be considered. These alterations are made using genetic operators and are most commonly mutation and/or crossover. The evaluation step assigns the objective value to each of the new candidate solutions by evaluating them on the problem (in this case the WDN problem). The selection step chooses which of the previous (parent) solutions and the new (child) solutions are kept for the next iteration to form the new set of parent solutions. This experiment varies the genetic operator in the variation step. The operators are described in a section below.

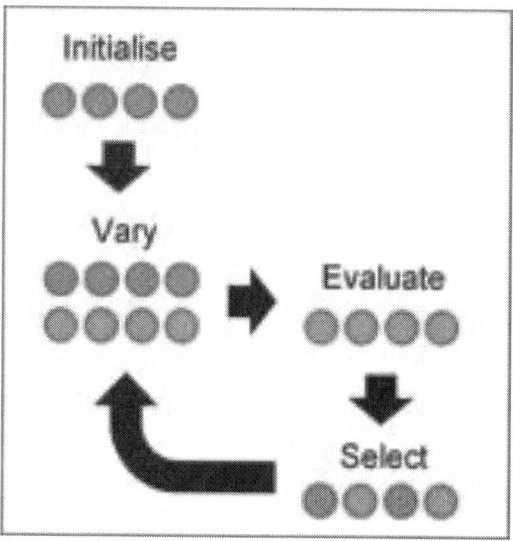

Figure 4. Illustration of a simple Evolutionary Algorithm (where circles represent solutions and black arrows the process flow).

The EA used in this study used an elitist selection strategy, ranking all of the parent and child solutions and selecting the best half (minimal cost). The parents selected for crossover operators were selected at random from the elitist parent population. The EA was given a population of 20 solutions (i.e., 20 parents to produce 20 children) and was run for 1000 generations on each problem (artificial and benchmark). The EAs were trialled 100 times on each problem and the median values were taken for the results to create a more fair comparison of results and reduce the impact of outlier results in the study.

Genetic operators

As explained above, the genetic operators in an EA provide the operational process for taking one or two existing solutions and generating new candidate solutions based on these pre-existing solutions. Mutation operators take one solution, for example, and perturb the existing parameter values to create a variation on the parent solution. Crossover, on the other hand, takes two parent solutions and swaps some of the parameter values between the two parents to create two new solutions which are recombined values from both parents. The latter being an emulation of the natural mating process. In addition to these traditional operators, this study explored operators designed specifically for the WDN design problem. Each of these six operators are described below.

Mutation (variants)

Two mutation operators were used that mutate only one pipe diameter selected at random from the solution's parameter vector. The mutation operators were: random and 1 step size variation. The random mutation replaced the existing pipe size with a uniformly random selected pipe diameter from the set of valid pipe diameters. The 1 step mutation increased or decreased the pipe diameter (chosen at random) by one pipe size. If the pipe was the smallest possible size it automatically increased it. Conversely, if it was the largest possible size it decreased it.

Crossover (variants)

The crossover operator selected two solutions at random and recombined their values to create a new solution. There were two variants of the crossover operator: n-point crossover and uniform crossover.

The n-point crossover placed a random number of crossover points in the parameter vector and swapped the parameter values between the odd and even points to create two new solutions. The number of crossover points was between 1 and one quarter of the parameter vector length (i.e., one quarter of the number of pipes in the network). The crossover points were selected at random.

The uniform crossover operator generated a mask of random Boolean values of the same length as the parameter vector. For every parameter position that contained a true value, the parameter values were swapped. By randomly swapping parameter values, this crossover operator generated two new solutions.

Pipe smoothing

The locally adaptive operator (inspired by the work of Johns et al., 2013) which has a pipe smoothing effect was the first of the two specialised operators for the WDN design problem. Firstly, this operator randomly mutated a pipe size using the 1 step mutation operator. Next, the pipe smoothing operator calculated which nodes had excess head (i.e., more pressure than necessary) and placed them in a list. It then selected one of these nodes at random and decreased the upstream pipe diameter by one size, to reduce the excess head at that node. This operator therefore provided a corrective operation to the 1 step mutation operator.

Pipe expander

The pipe expander operator (employing an opposing operation to the pipe smoothing operator) again took the 1 step mutation operator and applied an additional pipe expansion operation. The operator worked by selecting a pipe at random and, if the downstream node had a head deficit or head less than 40 m it would increase the pipe size by one step.

EXPERIMENTAL SETUP

Two experiments were conducted in this study where a standard EA was used to optimise two sets of networks – one set of artificial networks and one set of benchmark networks.

In the first experiment, shown in Fig. 3, a set of different genetic operators were tested, each embedded in individual EAs. The first experiment tested each of the operators on the artificial problems to determine which features, if any, had an impact on the efficacy of the operator for optimising the single objective cost minimization problem. After each of the operators were tested, the operators were tested on the benchmark problems and the results analysed in the context of the earlier results. A second experiment was then conducted which examined the effect of pairing operators. The pairs of operators were compared on artificial problems and the impact on each of the operator's performance was analysed.

Network simulation

The hydraulics for the networks were simulated using the well-known EPANet 2.0 hydraulics solver (Rossman, 2000). All the results were calculated using the full period simulation and the worst pressure values from the whole period taken for the objective value (i.e., the lowest head for each node over the whole period).

In this experiment, 60 networks were used which varied the presence of loops, dendritic branches, combined loops and dendritic branches (hybrid), single source, double sources (close), double sources (spread), valves, and pumps. Each combination of these features were used based on the three network structures, given in Fig. 5 below. The position for the optional elements (additional sources, valves and pumps) are also shown for reference.

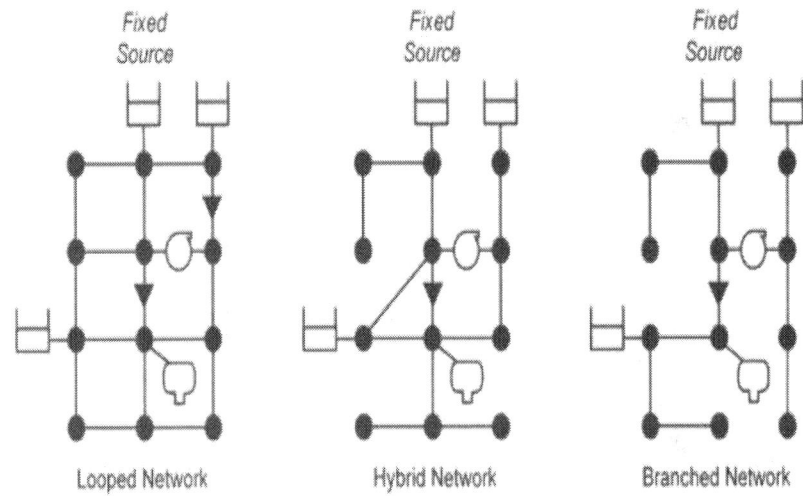

Figure 5. The three network structures used to create the artificial networks for testing different problem features. Open boxes represent reservoir sources, circles are junctions and demand nodes, the pump is shown by a circle with an outlet, the valves as arrows and the tank as a rounded rectangle with outlet. One source is fixed while all others are optional depending on the feature(s) needed in the artificial problem.

All the problems are constructed of 24 pipes with the same set of potential diameters and costs. The maximum total and minimum total costs for all of the artificial networks are the same, $600 to $1200 range. The diameters and associated costs per metre are given in Table 1.

For networks with pumps, the pumping costs were fixed at $50 per hour with schedules controlling the activation of pumps every 3 h allowing for a total of 8 changes to the pumping schedule in a 24 h period.

Benchmark problems

Three benchmark networks were used in the second experiment: Two Loops, Hanoi, and Anytown (Walters et al., 1999). These three networks can be downloaded from: http://emps.exeter.ac.uk/engineering/research/cws/resources/benchmarks/ and were selected as they represented well known benchmarks to compare the algorithms. Anytown presents the most complex network and is the only benchmark to include features such as pumps and tanks.

Parameter structure

The WDN problem is concerned with minimizing the cost of the network's construction with a constraint on valid values. The structure of the network, placement of sources, valves and pumps are fixed. Only the pipe diameters and, where applicable, pumping schedules could be altered in the network and as such are presented to the evolutionary algorithm as decision variables. A solution represented a list of the selected pipe size for each pipe in the network. This was represented as a vector of pipe diameter values, where the value at a given position related to a specific pipe in the network. I.e., a solution K with N pipes and 8 pump activation points is represented as $K = \{d1, d2, d3, ..., dn, p1, p2, ..., p8\}$. The diameters were stored as integer numbers representing the index of the pipe size (i.e., 0 = smallest pipe, 1 = second smallest pipe, and so on). The pipe activation values were stored as integers in the set $\{0, 1\}$ where 1 indicates the pump is on for the next 3 h period and 0 indicates the pump is off for the next 3 h period. Three hour periods were selected as it reduced the size of the parameter string and fitted well with the diurnal pattern of the network.

Experimental settings

The settings for the two experiments, including evolutionary algorithm parameters are given in Table 2 below.

Table 2. Experimental settings.

Setting	Experiment 1	Experiment 2
Number of Operators/Combinations	6	15
Artificial Networks	60	60
Trials per network	100	100
Generations	1000	1000
Population Size	20	20
Selection strategy	Elitist (truncate)	Elitist (truncate)
Archive Size	1 (best found)	1 (best found)

RESULTS

General performance

Fig. 6 shows the median final objective value obtained by each of the six operators on all of the 60 artificial problems. The problems are given on the x-axis and the median final objective value on the y-axis (to be minimized). (Note: all the artificial problems objective value ranges are the same). Each of the operators are indicated with the same symbol for each objective.

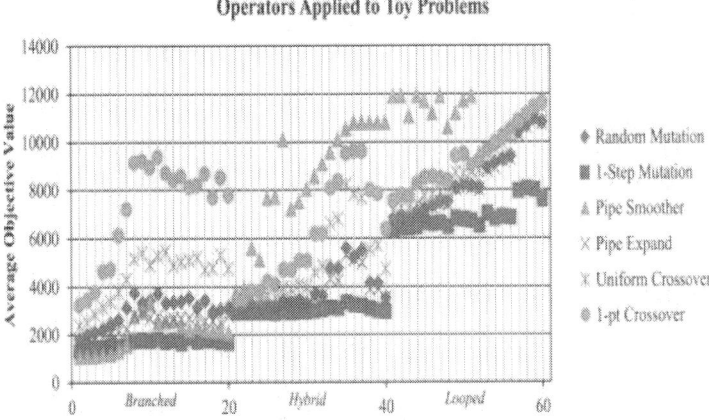

Figure 6. Plot of the median objective value obtained by each operator on each artificial problem. The x-axis relates to each problem and the y-axis the median objective value (averaged over all the trial runs). The first 20 problems are the branched networks, the second 20 the hybrid network and the last 20 the looped networks. In each set of 20, the first 10 networks do not use pumping, in contrast to the second 10, which do.

The 1-Step Mutation operator performs the best across all the problems with the exception of some of the branched networks. This is to be expected as the mutation operator does not employ any problem specific mechanisms or make large perturbations in the search space. The crossover operators perform poorly when applied alone and this can be explained by the fact that they make large perturbations in the search space which makes it more difficult for them to converge on good solutions as they are less capable of "fine tuning" existing good solutions. Additionally, the crossover operator is limited to the set of potential solution parameter values randomly generated at the start of the search as those operators do not introduce new parameter values but rather rearrange the existing values which limits the areas of the search space that can be searched.

The specialist operators do not have a fixed performance across all the problems. As is noted later, they are better at solving the branched networks (first 20 problems) than the looped networks. This is to be expected, especially for the pipe smoothing operator, which attempts to minimize excessive pipe diameters. The operator is based on the principle that the further downstream from the source the pipes are, the smaller the diameter is required as more of the demand is upstream (although only rudimentarily implemented). The pipe expander is comparable, in performance terms, with the random mutation operator whereas the pipe smoothing operator is very efficient at solving the simpler branched networks and obtains the best results across those problems with fewer problem features.

The introduction of the pumping schedule (the last 10 sets of results in each block of 20 results) can be seen to have a visible effect on the ability of the optimisers to converge on a good result. The pipe smoothing operator is visibly affected by this on the branched networks.

Convergence rates
To examine the convergence rates for each of the operators, Fig. 7 shows the number of generations taken for each operator to obtain their best objective value (averaged over all trials). As in Fig. 6, the x-axis displays each problem. The median number of generations to converge is given on the y-axis.

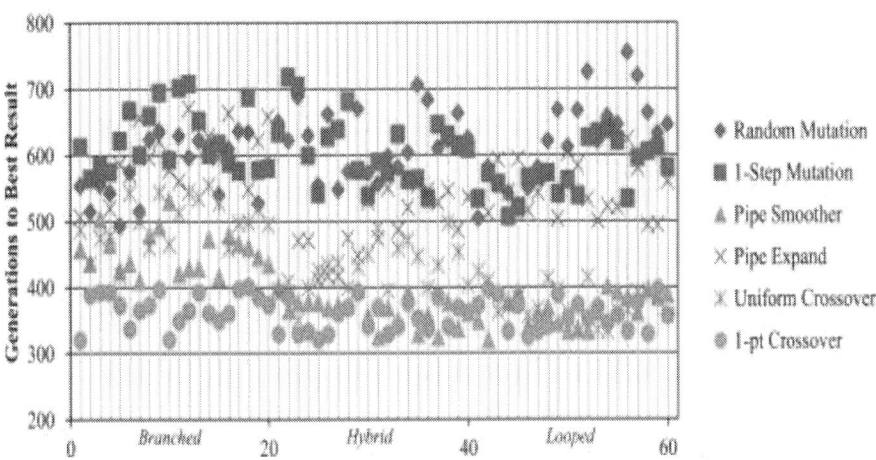

Figure 7. A plot showing the median number of generations for each operator to find their best objective value result (converge).

The standard mutation and 1-pt crossover operators converge at a similar rate across all the problems, although mutation operators converge more slowly than the crossover operator in general. The fast converging, relatively poorer results produced by the crossover operator suggests it converges early and is not able to effectively move about the search space. The mutation operators, however, are more explorative and hence converge more slowly and ultimately find better solutions than the crossover operator as is expected.

As expected, the pipe smoother converges quickly across the set of problems. Although, interestingly, it takes longer to converge on the branched networks. This is due to the continual fine tuning of the branched network to obtain the best possible network. The faster convergence on the hybrid and looped networks coupled with the worsened performance suggests the pipe smoother converges early on a poor result and is unable move away from the false optima it locates.

Loops and branches

As can be seen in Fig. 6, the predominant effect of introducing loops is to increase the final costs of the solutions discovered by all operators. In terms of relative performance among operators, the 1-step mutation remains fixed as the best performer and random mutation second across all problems. This indicates that the standard mutation operators are both robust methods for solving any WDN structure, which is expected as the operators do not employ any domain specific knowledge and only apply small perturbations to existing solutions. The presence of loops in the network did not affect the relative performance of crossover operators with respect to mutation. After examining the results produced by the crossover operator it is clear that the looped networks are more robust to large changes (due to the natural hydraulic robustness of the looped structure). However, as can be seen in Fig. 7, the looped networks are more slowly optimised by all the mutation and specialist operators (the first 20 problems).

As expected, the two specialist operators are more effective at solving the branched networks than those with loops and show a clear deterioration in performance when loops are introduced. However, the specialist operators are marginally more effective for solving the branched artificial problems than the standard mutation operators.

Multiple sources, pumps and valves

The presence of multiple sources does not seem to affect any of the operators in terms of convergence rates. Pumps, in contrast, significantly affect the final results for the majority of operators, particularly on the looped networks where most of the highest objective values are seen. This is not the case for the pipe smoothing operator though which shows worse performance on pumped hybrid networks, but slightly better performance on pumped looped networks. For hybrid and branched networks this can be explained as pumps interrupt the normal hydraulic conditions found in gravity fed systems and so disrupt the assumptive basis of the pipe smoothing operator (pipes further downstream generally need to supply less demand and so can be smaller).

The presence of valves (first four networks in each set of 20 results) improves the performance of the two specialist operators as it introduces flow restrictions on the looped networks which simplifies the network hydraulics. Furthermore, the presence of pumps in the looped network improves the performance of the pipe smoothing operator, as shown in Table 3. Here we can see that generally, the introduction of pumps reduces the performance of all the operators on all the networks with the exception of the pipe smoothing operator which receives a small improvement when pumps are present in the looped network. In effect, the pump provides a similar effect to the valves in these looped networks, restricting flow. The addition of the pumps increases the complexity of the problem and the length of the chromosome. Adding pumps introduces additional parameters to be optimised and makes the search space more complex. As such, the reduced performance of the operators within a fixed computational budget is to be expected.

Table 3. The averaged differential values between final objective function values depending on the presence of pumps in the network. Values below 1 indicate an overall reduction in performance when pumps are present in the network. Values greater than 1 indicate an improvement in performance.

Pump differential	Random mutation	1-Step mutation	Pipe smoother	Pipe expand	Uniform crossover	1-pt Crossover
Branched	0.81	0.89	0.76	0.83	0.75	0.72
Hybrid	0.72	0.94	0.65	0.74	0.58	0.53
Looped	0.77	0.91	1.07	0.77	0.77	0.81

Combinations of operators

After examining each operator in isolation, tests were conducted to explore the effect of combining pairs of operators. Table 4 below shows the results from these tests. Each combination of operators were tested on each artificial problem for 100 trial optimisation runs. The final population median objective value from each trial was compared with the median result obtained by each operator in isolation in the previous experiment. Table 4shows the count of trial runs that obtained better results than the median from the previous experiment.

Table 4. Frequency of improved results from combining operators over the sets of branched, hybrid and looped networks. Improvement relates to the operator given in each row when applied in conjunction with the operator assigned to each column. For example, the first row (Random Mutation) had improved results in 1815 trial runs out of 2000 when applied with the second row (1-Step mutation).

	Random Mutation	1-Step Mutation	Pipe Smoother	Pipe Expand	Uniform Crossover	1-pt Crossover
All Networks						
Random Mutation		62.15%	99.95%	55.53%	46.72%	41.77%
1-Step Mutation	30.25%		99.87%	36.00%	24.32%	19.07%
Pipe Smoother	95.83%	99.02%		97.52%	91.83%	87.77%
Pipe Expand	51.03%	62.32%	99.95%		46.10%	41.03%
Uniform Crossover	60.47%	68.95%	99.92%	62.92%		50.65%
1-pt Crossover	67.03%	76.47%	99.98%	70.18%	62.08%	
Branched Network						
Random Mutation		11.65%	99.90%	5.20%	0.80%	0.10%
1-Step Mutation	0.35%		99.65%	1.65%	0.10%	0.00%
Pipe Smoother	87.50%	97.05%		92.55%	75.50%	63.30%
Pipe Expand	1.55%	10.75%	99.90%		0.95%	0.15%
Uniform Crossover	5.60%	18.10%	99.75%	10.20%		0.95%
1-pt Crossover	14.10%	34.90%	99.95%	19.85%	6.55%	
Hybrid Network						
Random Mutation		76.85%	99.95%	65.65%	48.90%	38.80%
1-Step Mutation	19.80%		100.00%	27.90%	13.00%	7.15%
Pipe Smoother	100.00%	100.00%		100.00%	100.00%	100.00%
Pipe Expand	57.90%	77.95%	100.00%		47.85%	38.25%
Uniform Crossover	77.95%	89.70%	100.00%	80.65%		58.65%
1-pt Crossover	87.35%	94.70%	100.00%	91.20%	81.65%	
Looped Network						
Random Mutation		97.95%	100.00%	95.75%	90.45%	86.40%
1-Step Mutation	70.60%		99.95%	78.45%	59.85%	50.05%
Pipe Smoother	100.00%	100.00%		100.00%	100.00%	100.00%
Pipe Expand	93.65%	98.25%	99.95%		89.50%	84.70%
Uniform Crossover	97.85%	99.05%	100.00%	97.90%		92.35%
1-pt Crossover	99.65%	99.80%	100.00%	99.50%	98.05%	

The better than average results for operator is given in each row. For example, the row for Random Mutation shows the number of better than average results obtained by that operator when applied in conjunction with each operator assigned to each column. I.e., Random Mutation and 1-Step Mutation together obtained 62.15% better than average Random Mutation alone when applied on all networks. From this it can be observed that, across the range of problems, Random Mutation performed better on average when paired with 1-Step mutation.

The results for all networks show the better than average results from a total of 6000 trial runs (accumulated across all the problems). The results for each of the three types of networks show the percentage of better than average results over an accumulated 2000 trial runs from all the networks of that type. A result of over 50% indicates the operator performs better when paired with the counter-part operator. Less than 50% indicates the operator is hindered by the other operator.

Based on the results in the previous experiment, we can see that all the operators were generally improved when applied in conjunction with 1-Step Mutation and the Pipe Smoother operator. Interestingly, 1-Step mutation was not generally improved by the addition of any other operator; with the exception of the Pipe Smoothing operator. This is an interesting result as it shows that the 1-Step Mutation was not (across the whole set of problems) improved by the addition of a crossover operator as might be expected. In contrast, the crossover operators benefitted from the addition of every other type of operator; especially the 1-pt Crossover.

The Pipe Smoothing operator is a particularly interesting case as the results highlight the "tuning" behaviour of the operator. The random perturbations created by the other operators are then "fixed" or "enhanced" by the Pipe Smoother and so it boosts the performance of these more stochastic operators. Likewise, the Pipe Smoothing operator greatly benefits from random disruptions to its early converging search behaviour and is able to better search the solution space.

When these results are examined more closely it is clear that all the stochastic operators perform better on the branched networks when applied in isolation whereas all the operators were improved by pairing when applied to the looped networks. The branched network has weaker parameter interdependencies and is therefore a simpler search space and set of problems to solve. The looped network is more difficult to solve as the loops create stronger hydraulic interdependencies between the pipes and so presents a more difficult problem. The 1-Step mutation is the only operator that has the least enhancement when applied with every operator. For example, it is not significantly improved by the addition of the 1-pt Crossover. Indeed, on the hybrid network, the 1-pt Mutation is disadvantaged by the addition of every other operator (with the exception of Pipe Smoothing). These results indicate that the generally better performance of the 1-Step Mutation is maintained, even when compared with pairings of other operators.

It is interesting to note that the 1-pt Crossover operator performs poorly across all the problems and, as shown in Table 4, provides the least improvement when applied in conjunction with other operators. This suggests that this operator is not well suited to this class of problem. Indeed, the 1-pt crossover operator is dependent upon the length and encoding type used by the problem and its application may not be well suited to the structure of this problem; as indicated by the results. That is not to say it is a bad operator, rather that it is not well suited to the problems examined here.

Table 5 and Table 6 below show the median result obtained by each combination of operators on the Branched_1 and Looped_1 networks. These networks represent the 1st network of the branched and looped topologies with gravity fed systems, without pumps or tanks. In both cases the 1-Step Mutation with Pipe Smoother operator obtains the best average value compared to the other pairs. These two example sets of results are indicative of the majority of results found across all the problems.

Table 5. Median objective value obtained by each pairing of operator on the Branched_1 network, averaged over 100 trial runs. The best objective value is highlighted in bold.

Network: Branched_1	1-Step Mutation	Pipe Smoother	Pipe Expand	Uniform Crossover	1-pt Crossover
Random Mutation	3488.24	2884.102	3788.862	3889.922	4362.794
1-Step Mutation		**2811.675**	3260.32	3516.47	3859.401
Pipe Smoother			2850.433	2983.779	3196.36
Pipe Expand				4002.19	4214.962
Uniform Crossover					4720.72
1-pt Crossover					

Table 6. Median objective value obtained by each pairing of operator on the Looped_1 network, averaged over 100 trial runs. The best objective value is highlighted in bold.

Network: Looped_1	1-Step Mutation	Pipe Smoother	Pipe Expand	Uniform Crossover	1-pt Crossover
Random Mutation	6492.854	6124.922	6574.066	6835.196	6950.607
1-Step Mutation		**6003.437**	6410.029	6554.535	6664.378
Pipe Smoother			6060.344	6206.595	6315.911
Pipe Expand				6657.225	6818.499
Uniform Crossover					7059.035

The results show in Table 5 and Table 6 demonstrate how different operators can be combined to beneficial (and detrimental) effect. Additionally, the combination of the random mutations provided by the 1-Step Mutation with the converging and problem specific action of the Pipe Smoother is a clear example of how two behaviours can be applied together in a complementary way to improve the search.

Benchmark problems

This section examines the results from the second experiment on the benchmark networks. Fig. 8 shows the median objective value at the final generation for the six operators on the three benchmark networks. The assumed performance (prior to the experiment) are shown below followed by an analysis of their performance.

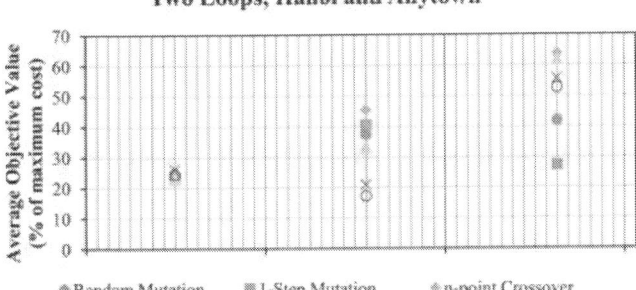

Figure 8. The median of the final objective value obtained by each operator on the three benchmark networks. Results are shown as a percentage of the maximum cost for that network to normalise across the three networks to allow for comparison of results.

Two loops

The two loops system contains loops but no pumps and tanks and is most similar to Network 14 from the artificial problems. All the operators are expected to perform reasonably well on this network, although the two specialist operators are not expected to perform better than the normal mutation operators. As can be seen by the results, all the operators perform as expected, with the two crossover operators performing better than expected. The Two Loops network is simpler than the artificial networks and so accounts for this improved performance as the search space is smaller and the disruptive effect of the crossover operator is relatively lower compared to the single point mutation operators.

Hanoi

The Hanoi network contains loops and no dendritic branches. The larger number of pipes suggests the highly disruptive crossover operators will perform less well compared to the mutation operators and specialist operators due to the greater number of perturbations to the solution that will occur as a result of crossover. The gravity fed nature of this network actually produces an almost branched hydraulic system, as the pipes further away from the source need to feed only the demand of those nodes downstream. Due to the single source and this effect, there is often no requirement for flow to circulate around a loop, creating the 'branched' effect. This effective branching suggests the specialist operators will perform better on this problem.

As is shown by the results in Fig. 8, the specialist operators perform the best on this problem. These operators combine the explorative power of the mutation operator and also the faster converging properties of the functions that modify the network in line with its hydraulic properties (e.g. pipe smoothing and pipe expand). The pipe smoothing operator, in particular, works very effectively on this network. As expected, the crossover operators are not able to optimise this problem in isolation. The large parameter string prevents the large perturbations of the crossover operator from converging on a solution. The mutation operators perform reasonably well again on this problem, although convergence is slower than the specialist operators as expected.

Anytown

This network is the most complex of the three networks tested and highly looped with the presence of additional tanks and pumps (which were scheduled as part of the optimisation). The specialist operators are not expected to perform particularly well on this problem and, similarly, the crossover operator is also not expected to perform well.

As can be seen in the results (see Fig. 8), the crossover operator is not able to effectively optimise this problem. The two specialist operators perform better than expected but converge early. The two standard mutation operators are slower to converge but produce better results by the final generation. Again, this network demonstrates how the assessment of the operators on the artificial problems enables an effective prediction of performance prior to optimising the larger, real-world or benchmark networks.

CONCLUSIONS

This paper has presented and given a practical demonstration of a method for assessing Evolutionary Algorithm operator behaviour in the context of specific WDN design problem features, such as pumps, tanks and loops. The method isolates the structural features of a WDN network and systematically assesses whether any correlation occurs between an operator and a feature. This method provides a means for establishing a prior understanding of an operator's efficacy on different WDN networks and aids in the development of new specialist operators and the selection of the most appropriate operators for a given network.

The paper demonstrates the method by testing six EA operators on 60 artificial networks that contain specific features. The performance of the six operators are observed and analysed and show how the more general 1-Step mutation operator performs well across the different artificial networks in comparison to the operators specialised for specific topologies (dendritic systems).

A similar comparison of the combination of operators is then conducted and the effect of combining different operators analysed. The results demonstrated how combining operators can be effective in improving EA search results in some situations. However, it is also shown that the mutation operators are more effective at solving the WDN problem on the simpler dendritic systems compared to the looped systems where combined operators are more effective. Overall, the combination of the 1-Step mutation operator and Pipe Smoothing operator is shown to be the most beneficial combination of operators.

The results presented in this paper provide evidence for the idea that operator performance and problem search spaces are linked in water distribution network optimisation, a notion that has been shown to be true in other problem spaces.

ACKNOWLEDGEMENTS

This work was funded by the UK Engineering and Physical Sciences Research councilunder grant number EP/K000519/1.

REFERENCES

1. Afshar, M., 2006. Improving the efficiency of ant algorithms using adaptive refinement: application to storm water network design. Adv. Water Resour. 29 (9), 1371e1382.
2. Bi, W., Dandy, G.C., Maier, H.R., 2015. Improved genetic algorithm optimization of water distribution system design by incorporating domain knowledge. Environ. Modell. Softw 69, 370e381. http://dx.doi.org/10.1016/j.envsoft.2014.09.010.
3. Blum, C., Roli, A., 2003. Metaheuristics in combinatorial optimization: overview and conceptual comparison. ACM Comput. Surv. 35 (3), 268e308.
4. Burke, E., Gendreau, M., Hyde, M., Kendall, G., Ochoa, G., Ozca, E., Qu, R., 2013. Hyper-heuristics: a survey of the state of the art. J. Oper. Res. Soc. 64 (12), 1695e1724.
5. Cheung, P.B., Reis, L.F., Formiga, K.T., Chaudhry, F.H., Ticona, W.G., 2003. Multiobjective evolutionary algorithms applied to the rehabilitation of a water distribution system: a comparative study. In: Evolutionary Multi-criterion Optimization, Lecture Notes in Computer Science, vol. 2632. Springer Verlag, pp. 662e676.
6. Coello, C.C., Lamont, G.B., Van Veldhuizen, D.A., 2007. Evolutionary Algorithms for Solving Multi-objective Problems, second ed. Springer, New York.
7. Deb, K., Jain, S., 2002. Running Performance Metrics for Evolutionary Multiobjective Optimization. Indian Institute of Technology, Kanpur.
8. Franchini, M., Galeati, G., 1997. Comparing several genetic algorithm schemes for the calibration of conceptual rainfall-runoff models. Hydrol. Sci. J. 42 (3), 337e379.
9. Fu, G., Khu, S.-T., Butler, D., 2008. Multiobjective optimisation of urban wastewater systems using ParEGO: a comparison with NSGA II. In: 11th International Conference on Urban Drainage, Edinburgh, Scotland, UK.

10. Hansen, N., et al., 2010. Comparing results of 31 algorithms from the blackbox optimization benchmarking BBOB-2009. In: Workshop Proceedings of the GECCO Genetic and Evolutionary Computation Conference 2010. ACM.
11. Jha, M.K., Nanda, G., Samuel, M.P., 2004. Determining hydraulic characteristics of production wells using genetic algorithm. Water Resour. Manag. 18 (4), 353e377.
12. Johns, M.B., Keedwell, E., Savic, D.A., 2013. Pipe smoothing genetic algorithm for least cost water distribution network design. In: GECCO 2013, pp. 1309e1316.
13. Kauffman, S.A., 1989. Adaptation on rugged fitness landscapes. In: Lectures in the Sciences of Complexity. Addison-Wesley Longman, Amsterdam, pp. 527e618.
14. Keedwell, E., Khu, S.T., 2005. A hybrid genetic algorithm for the design of water distribution networks. Eng. Appl. Artif. Intell. 18 (4), 461e472.
15. Laumanns, M., Zitzler, E., Thiele, L., 2000. A unified model for multi-objective evolutionary algorithms with elitism. In: Congress on Evolutionary Computation (CEC 2000), La Jolla, CA, pp. 46e53.
16. Maier, H.R., Kapelan, Z., Kasprzyk, J., Kollat, J., Matott, L.S., Cunha, M.C., Dandy, G.C., Gibbs, M.S., Keedwell, E., Marchi, A., Ostfeld, A., Savic, D., Solomatine, D.P., Vrugt, J.A., Zecchin, A.C., Minsker, B.S., Barbour, E.J., Kuczera, G., Pasha, F., Castelletti, A., Giuliani, M., Reed, P.M., 2014. Evolutionary algorithms and other metaheuristics in water resources: Current status, research challenges and future directions. Environ. Model. Softw. 62, 271e299. http://dx.doi.org/10.1016/j.envsoft.2014.09.013.
17. Malan, K.M., Engelbrecht, A.P., 2013. A survey of techniques for characterising fitness landscapes and some possible ways forward. Inf. Sci. 241, 148e163. http://dx.doi.org/10.1016/j.ins.2013.04.015.
18. Marchi, A., Dandy, G.C., Wilkins, A., Rohrlach, H., 2014. Methodology for comparing evolutionary algorithms for optimization of water distribution systems. J. Water Resour. Plan. Manag. 140 (1), 22e31.
19. McClymont, K., Keedwell, E., Savic, D., Randall-Smith, M., 2013. A general multiobjective hyper-heuristic for water distribution network design with discolouration risk. J. Hydroinform. 15 (3), 700e716. http://dx.doi.org/10.2166/ hydro.2012.022.
20. Merz, P., 2004. Advanced fitness landscape analysis and the performance of memetic algorithms. Evol. Comput. 12 (3), 303e325.
21. Moraglio, A., Poli, R., 2004. Topological interpretation of crossover. In: GECCO 2004, Lecture Notes in Computer Science 3102. Springer, pp. 1377e1388.

22. Nicklow, J.W., Ozkurt, O., Bringer Jr., J., 2003. Control of channel bed morphology in large-scale river networks using a genetic algorithm. Water Resour. Manag. 17 (2), 113e132.
23. Piscopo, A.N., Kasprzyk, J.R., Neupauer, R.M., 2014. An iterative approach to multiobjective engineering design: optimization of engineered injection and extraction for enhanced groundwater remediation. Environ. Model. Softw. Available online 30 September 2014.
24. Raad, D., Sinske, A., Vuuren, J.v., 2010. Multiobjective optimization for water distribution system design using a hyperheuristic. J. Water Resour. Plan. Manag. 592.
25. Rossman, L.A., 2000. EPANET Users Manual. Technical Report No. EPA/600/R-00/ 057, United States 710 Environment Protection Agency.
26. Savic, D.A., Walters, G.A., 1995. An evolution program for optimal pressure regulation in water distribution networks. Eng. Optim. 24 (3), 197e219.
27. Savic, D., Walters, G., 1997. Genetic algorithms for least-cost design of water distribution networks. J. Water Resour. Plann. Manag. 123 (2), 67e77.
28. Simpson, A.R., Dandy, G.C., Murphy, L.J., 1994. Genetic algorithms compared to other techniques for pipe optimization. J. Water Resour. Plan. Manag. 120 (4), 423e443.
29. Van Veldhuizen, D.A., 1999. Multiobjective Evolutionary Algorithms: Classifications, Analyses, and New Innovations (PhD thesis). Department of Electrical and Computer Engineering. Graduate School of Engineering, Air Force Institute of Technology, Wright-Patterson AFB, Ohio.
30. Van Veldhuizen, D.A., Lamont, G.B., 1998. Evolutionary computation and convergence to a pareto front. In: Koza, J.R., Banzhaf, W., Chellapilla, K., Deb, K., Dorigo, M., Fogel, D.B. (Eds.), Genetic Programming 1998: Proceedings of the Third Annual Conference. Morgan Kaufmann, pp. 221e228.
31. Vrugt, J.A., Robinson, B.A., 2007. Improved evolutionary optimization from genetically adaptive multimethod search. Proc. Natl. Acad. Sci. U. S. A. 708e711. http:// dx.doi.org/10.1073/pnas.0610471104.
32. Vrugt, J.A., Robinson, B.A., Hyman, J.M., 2009. Self-adaptive Multimethod Search for Global Optimization in Real-parameter Spaces. IEEE.
33. Walters, G.A., Halhal, D., Savic, D., Ouazar, D., 1999. Improved design of "Anytown" distribution network using structured messy genetic algorithms. Urban Water 1 (1), 23e38.
34. Walters, G.A., Lohbeck, T., 1993. Optimal layout of tree networks using genetic algorithms. Eng. Optim. 22 (1), 27e48.
35. Whitley, D., Rana, S., Dzubera, J., Mathias, K.E., 1996. Evaluating evolutionary algorithms. Artif. Intell. 85 (1), 245e276.

36. Wolpert, D.H., Macready, W.G., 1997. No free lunch theorems for optimization. IEEE Trans. Evol. Comput. 1 (1), 67e82.
37. Wright, S., 1932. The roles of mutation, inbreeding, crossbreeding and selection in evolution. In: Jones, D. (Ed.), Proceedings of the Sixth International Congress on Genetics, pp. 356e366.
38. Yates, D., Templeman, A., Boffey, T., 1984. The computational complexity of the problem of determining least capital cost designs for water supply networks. Eng. Optim. 7 (2), 143e155.
39. Zecchin, A.C., Simpson, A.R., Maier, H.R., Marchi, A., Nixon, J.B., 2012. Improved understanding of the searching behavior of ant colony optimization algorithms applied to the water distribution design problem. Water Resour. Res. 48, W09505.
40. Zitzler, E., Deb, K., Thiele, L., 2000. Comparison of multiobjective evolutionary algorithms: empirical results. Evol. Comput. 8 (2), 173e195.

CITATION

K. McClymont, E. Keedwell, D. Savic, An analysis of the interface between evolutionary algorithm operators and problem features for water resources problems. A case study in water distribution network design, Environmental Modelling & Software, Volume 69, July 2015, Pages 414-424, ISSN 1364-8152, http://dx.doi.org/10.1016/j.envsoft.2014.12.023.

CHAPTER 4

Vadose Zone Heterogeneity Effect on Unsaturated Water Flow Modeling at Meso-Scale

Artur Paiva Coutinho[1,2]*, Laurent Lassabatere*[1]*, Thierry Winiarski*[1]*, Jaime Joaquim da Silva Pereira Cabral*[2]*, Antonio Celso Dantas Antonino*[2]*, Rafael Angulo-Jaramillo*[1]

[1]LEHNA, UMR 5023 CNRS, ENTPE, UCB-Lyon-1, Vaulx-en-Velin, France
[2]Universidade Federal de Pernambuco, Centro de Tecnologia e Geociências, Programa de Pós Graduação em Engenharia Civil/Tecnologia Ambiental e Recursos Hídricos, Recife, Brasil

ABSTRACT

The understanding of unsaturated flow in heterogeneous formations is a prerequisite to the understanding of pollutant transfer in the vadose zone and the proper management of infiltration basins settled over such heterogeneous formations. This study addresses the effect of lithological heterogeneity of a glaciofluvial deposit on flow in the vadose zone underneath an infiltration basin settled in the Lyon suburbs. The basin had already been the subject of several previous studies, some of which demonstrated the impact of soil heterogeneity. But all of them were only based on the sedimentological study of a trench and no study addressed the potential spatial variability of results due to the spatial variability of soil heterogeneity. In this study, we model flow in the vadose zone for several case studies, including drainage, water infiltration during a rainfall event, and a complete meteorological chronic. These calculations were conducted for several sections, previously characterized in the basin using GPR and sedimentological study and compared with a blank (homogeneous section). The results clearly show that heterogeneity impacts unsaturated flow and that these impacts depend upon the section considered. Some geometrical architectural and textural parameters were proposed to explain the spatial variability and effect of the soil heterogeneity on unsaturated flow, thus establishing the first step towards modeling unsaturated flow in the basin at the meso-scale.

INTRODUCTION

Nowadays cities expand as a result of urbanization. The extension of urbanization and sealed surface results in the modification of the water cycle, and the volumes produced by the sealed surface are too important to be collected and treated in usual stormwater plants. To alleviate such problems, best management practices have been developed, most of which are based on the infiltration of runoff water in infiltration basins or pits. Infiltration basins must be settled over permeable soils to allow water infiltration. Recently, the effects of infiltration basin on water cycle and on the quality of groundwater and the soil below have been questioned [1] . A particular care must be taken about the pollutants carried by runoff water (heavy metals, organics, etc.) that can be responsible for the degradation of the quality of the soils.

In the east plain of Lyon (France), infiltration basins are commonly used to infiltrate runoff water. In this region, most infiltration basins were settled over a glaciofluvial deposit, as a result of ancient glaciation periods and the action of the Rhone River. This glaciofluvial deposit has proved to be highly heterogeneous and made of contrasting material in terms of sedimentological properties. Goutaland et al. [2] [3] and Winiarski et al. [4] have proposed a methodology to assess flow and pollutant transfer in such heterogeneous media. The first step consists of describing the sedimentological properties of the heterogeneous media using geophysical tools. Ground penetrating radar (GPR) can be used as a non-destructive tool for investigating glaciofluvial deposits in detail up to a depth of 20 m [5] [6] . GPR allows the documentation of the internal structure (bedding geometry) of active braided bars and channels under both saturated and unsaturated conditions [6] [7] . Using GPR, Goutaland et al. [3] proposed a methodology to identify the different units and materials, referred to as lithofacies, which constituted the deposit and dug a trench to validate their findings. Then, the same authors and colleagues used water infiltration experiments and best method [8] to derive the hydraulic properties of the lithofacies and to model drainage in the trench. On this basis, Winiarski et al. [4] coupled the modeling of flow with the hydrodispersive and geochemical properties of the lithofacies to predict pollutant transfer over the same section. These authors concluded that preferential flow was likely to establish, in particular under unsaturated conditions with capillary barrier induced funneled flows. They also concluded that preferential flow might impact significantly the transfer of pollutants and reduce the removal of pollutants by materials that were far from flow pathways. Indeed, it is well known that pollutant transfer in the soil is highly sensitive to flow pathways and flow heterogeneity since preferential flow may reduce the access of pollutants to reactive particles, e.g. [9] -[11].

In these studies, the investigations were focused on the case of one sole section, for which the sedimentological information was validated with a trench. For this section, flow was modeled for several scenarios in terms of geometry (from a homogeneous section filled only with the main material to the most heterogeneous case with a full description including all lithofacies). If the results show that the heterogeneity plays a role, this must be confirmed with the study of several sections and scenarios. Indeed, only one section was investigated over an infiltration basin of 1.38 ha. To get an idea of the impact of preferential flow on the whole infiltration basin, we must get more information on the spatial variability of heterogeneity of sedimentological properties of the deposit and the effect of such variability on preferential flow.

In this study, we focus on the impact of sedimentological heterogeneity on unsaturated flow. For that purpose, we have scanned the ground with GPR over several sections and we have modeled the flow in these different sections and compared the results with the flow in a homogeneous section containing the main material. Several hydric and hydraulic conditions were tested: drainage, water infiltration (with fixed water flux) and a complete meteorological chronic of the first 1000 hours of 2008. We aim to understand and characterize the effect of soil heterogeneity on flow in the vadose zone, and in particular its effect on flow pathways and fluxes at surface for several kinds of heterogeneous soil profiles. In addition to the analysis of water pressure and fluxes in the profile and at surface, quantitative indicators are proposed to link flow pattern to the lithological heterogeneity of studied sections. These indicators are designed to summarize the architecture of the heterogeneous sections and are expected to serve as explanatory factors for flow pattern, with the final aim to be used as tools for the prediction of flow response to lithological heterogeneity. This is the first step in understanding and modeling of flow heterogeneity at the basin scale.

MATERIALS AND METHODS

Site location
The field site is a storm water infiltration basin, referred to as Django Reinhardt basin (DjR basin), and located in Chassieu, in the eastern suburbs of Lyon, France (Figure 1a)). The storm water catchment is an industrial area of 185 ha located south of Chassieu (Figure 1b)). The basin

covers an area of 1.38 ha and is downstream from a storage and settling basin. This basin is located on quaternary deposits of a glaciofluvial corridor (southeast?northwest orientation), deposited during the last glacial maximum (Figure 1a)). The thickness of the deposit is approximately 30 to 35 m (Barraud et al., 2002). The deposit rests on an impervious substratum of tertiary mollassic sands. The groundwater level is at a depth of 13 m. The aquifer has a mean hydraulic conductivity of 7 to 9×10^{-3} m s^{-1} [12]. This site is instrumented by the Observatoire de Terrain en Hydrologie Urbaine (http://www.graie.org/othu/).

GPR Study

Ground penetrating radar (GPR) is a non-invasive geophysical technique that detects electrical discontinuities in the shallow subsurface. This technique is based on the generation, transmission, propagation, reflection and reception of discrete pulses of high frequency (MHz) electromagnetic energy. It allows the assessment of spatial structures of sediments to a depth of 10 to 20 m over extended areas [13] -[15]. Sedimentary geologists widely used this technique to reconstruct depositional environments and document historic sedimentary processes [16]. Indeed, glaciofluvial deposits are characterized by a textural heterogeneity derived from erosion, transport and sedimentation phases that have led to the formation of lithological units. In the present paper, the term lithofacies has been used as reference to lithological units of glaciofluvial deposits, all of these corresponding to a specific genetic unit (i.e., formed by a homogeneous process of transport and sedimentation, [17]).

The GPR measurements were carried out using the GSSI SIR 3000 system (Geophysical Survey System Inc., Salem, USA), operated with a shielded antenna at a central frequency of 400 MHz and 200 MHz, running in monostatic mode. The data processing was performed using the GSSI Radan 7 software. This processing consisted of distance normalization, a static time shift (to align direct ground wave arrival to 0 ns), a background removal (to eliminate the high amplitude direct ground wave), and a Kirchhoff migration. The electromagnetic wave velocity was determined by CMP (common midpoint) with two 400 MHz antennas. According to Goutaland et al. [18], a velocity of 0.09 m ns^{-1} was selected to convert two-way travel time into actual depth.

MATERIALS AND METHODS

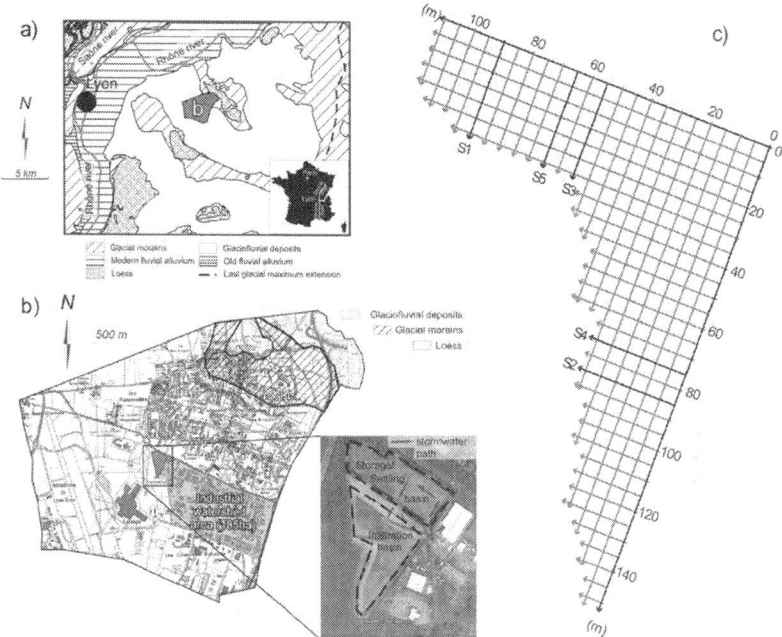

Figure 1. a) Geological settings of the eastern part of the Lyon area; b) location of site in the Chassieu city area. The DjR infiltration basin was settled over a 13-m-deep unsaturated glaciofluvial deposits and receive the waters from a storage and settling basin; c) Layout of the 5 m × 5 m rectilinear GPR acquisition grid that covers the infiltration basin.

At our site, previous GPR investigations at 200, 400, and 900 MHz had been performed and compared to trench walls [18] [19] . These data were used to explore the ability to detect sedimentary structure at the scale of the required model resolution (i.e., at the lithofacies scale) and the penetration depth for each antenna. The best tradeoff between high resolution and adequate penetration depth was obtained with the 200 MHz antenna. Thus, only results obtained using the 200 MHz antenna are illustrated.

The 400 MHz and 200 MHz profiles were collected in a 5 m × 5 m grid pattern (Figure 1c)). The dimensions of the acquisition grid were 140 m × 110 m. The spacing between each grid line was set at 5 m. GPR measurements (400 MHz and 200 MHz) were performed on all acquisition grid lines. GPR data set were interpreted using radar facies analysis [18] in conjunction with the classification of seismic reflections developed by Mitchum et al. [20] . During interpretation of the profiles, constant gain was applied to the data to note the amplitude of reflections. The GPR profiles were exported into image file format (.bmp), and major reflections were traced out and colored using Illustrator CS6 drafting software (Adobe, Inc.). As GPR profiles acquired an immense amount of interpreted sedimentological information, only one major radar reflection package was described with the profiles of 200 MHz antennae. The profiles of 400 MHz antennae (not presented in this paper) were used to confirm the previous interpretations based on the 200 MHz antennae.

Numerical modeling offlow

Unsaturated flow was modeled for several sections, considering the first 4 meters of soil below several lines of the acquisition grid. Indeed, each acquisition line was interpreted and GPR data was precise enough to describe the deposit architecture in detail (i.e., lithofacies and relative positions) for ~4 meters. For each section (i.e., ~4 m of soil below the acquisition line), unsaturated flow in the deposit was modeled considering Richard's equation and using HYDRUS 3D software [21] :

$$\frac{\partial \theta}{\partial t} = \frac{\partial}{\partial x}\left(K(\theta)\frac{\partial h}{\partial x}\right) + \frac{\partial}{\partial y}\left(K(\theta)\frac{\partial h}{\partial y}\right) + \frac{\partial}{\partial z}\left(K(\theta)\left(\frac{\partial h}{\partial z}+1\right)\right) \quad (1)$$

where θ is the volumetric water content, h is the pressure head, x, y, and z are the spatial coordinates, t is time, and K is the unsaturated hydraulic conductivity tensor. We assumed isotropy for all lithofacies. Soil water retention and hydraulic curves are characterized by the van Genuchten-Mualem relation [22] :

$$\theta(h) = \theta_r + (\theta_s - \theta_r)\left(1 + |\alpha h|^n\right)^{-m} \quad (2a)$$

$$K(h) = K_s K_r(h) = K_s S_e^l \left(1 - \left(1 - S_e^{1/m}\right)^m\right)^2 \quad (2b)$$

where θ_r and θ_s denote the residual and saturated water content, respectively; K_s and K_r are the saturated and relative hydraulic conductivity, respectively h_g the scale parameter of water pressure heads, n is a pore-size distribution index, l is a pore-connectivity parameter, and a shape parameter m equal to 1-1/n. On the basis of the sedimentology description obtained with the method described above, each material was assigned the hydraulic properties proposed by Goutaland [2] . Hydraulic parameters are described in Table 1. In this study, the role of the coarser material (gravel) was neglected and this topic will be the subject of further studies.

For numerical resolution, sections were meshed through triangular elements to build the reference 2D numerical domain, 30 m in length and 4 m in depth. Mesh elements dimensions never exceed 8cm and the number of nodes is in the order of 30,000 nodes. Each node was assigned a set of hydraulic accordingly to the sedimentological description of the trench. Boundary conditions were set at no flux on the side walls and free drainage at the bottom. Two events were modeled: drainage from a close to saturation initial state and water infiltration induced by fixed constant flux at surface, mimicking a rainfall event, using atmospheric boundary option. The initial condition was fixed at hi = −0.01 m for the drainage to simulate close to initial saturated condition and to −5 m for the rainfall event. In this study, the accumulation of water at soil surface was neglected. The maximum al-

Table 1. Hydrodynamic properties of the litohfacies.

Litofacies	Θr (cm^3/cm^3)	Θs (cm^3/cm^3)	hg (m)	n	Ks (m/s)
Gcm	0.0107	0.274	0.1562	1.7353	$1.53 \cdot 10^{-4}$
Gcm,b	0.0051	0.186	0.2299	1.9336	$1.91 \cdot 10^{-4}$
S-x	0.013	0.359	0.3559	1.7878	$5.1 \cdot 10^{-5}$

lowed pressure head at the soil surface, that allows the calculation of the real water flux at the atmospheric boundary, was fixed at its default value, i.e., 0. Finally, modeling was conducted for real meteorological data recorded in 2008 at the station of Lyon Bon (France). Incoming flow rates (q_{enter}) were calculated using rainfall and potential evaporation-transpiration intensities (i_{rain} and i_{pet}), the catchment runoff coefficient (c_R) and both catchment and basin surface areas (S_C and S_B), through: [23]

$$q_{enter} = \begin{cases} \dfrac{(c_R S_C + S_B)}{S_B} i_{rain} & \text{if rain} \\ i_{pet} & \text{else} \end{cases} \quad (3)$$

The runoff coefficient is fixed at a value of 0.4 and evaporation-transpiration was uniformly distributed all along the year.

RESULTS AND DISCUSSION

Architectural characterization of the Studied Sections
A sedimentological description of glaciofluvial deposit was performed at both a textural and structural scale. The structural scale is the description scale of the sedimentary bodies, composed of distinctive assemblages of lithofacies and corresponding to the depositional product of a particular process or succession of processes occurring during a depositional system [17] . These sedimentary bodies are called architectural units or depositional units [24] . The textural scale aims at characterizing lithofacies that are defined as uniform strata characterized by their distinctive lithological features (composition, grain size, bedding characteristics, and sedimentary structures) and corresponding to an individual depositional event [17] .

Both lithofacies and depositional units of the studied glaciofluvial deposit have already been characterized by Goutaland et al. [18] for the studied deposit. These authors described the lithofacies using the sedimentological code of Miall [17] extended by Heinz et al. [24] . According to results obtained for a trench studied by Goutaland et al. [19] in the same infiltration basin, we interpreted six of the radar sections of the acquisition grid depicted in Figure 1 (Sections S1 to S6).Figure 2 shows one example of the radar profile interpretation (S1). The proposed interpretation was conducted considering the six depositional units proposed by Goutaland et al. [18] , namely units 1, 2, 3, 4, 5 and 6, from the base to the top of each section (Figure 2a)).

Two distinct levels were revealed over the first four meters of the glaciofluvial deposit (Figure 2b)):

- An upper level (first ~0.5 m) corresponding to particle deposition at high stream energy, which lies at the origin of the presence of a very heterometric sand and gravel mix (Gcm lithofacies). This level comprises structural unit 6;
- A lower level (located at a depth > ~0.5 m, on average) that corresponds to a braided proglacial system deposit, leading to such structures as paleochannel and scour fill (S-x, Gcm and Gcm, b lithofacies), and gravel progradations (Gcm, b lithofacies). This level comprises the structural units 1 - 5.

In these last units, three distinct lithofacies are present [18] : medium sands (S-x lithofacies), sand and gravel mixes, with a dominant heterometric gravel fraction (Gcm lithofacies) or with a bimodal grain size distribution (Gcm, b lithofacies). The sand and gravel mixes (Gcm, b lithofacies) constitute the predominant lithofacies of the glaciofluvial deposit. Goutaland et al. [19] provided a sedimentological interpretation of all lithofacies. The sands are low-flow regime deposits, whereas the Gcm, (b) lithofacies correspond to a high flow stage deposit. The Gcm, b lithofacies formed by gravel dunes migration resulted from gravel bars migration. These considerations allow us to define the edges of all lithofaciesin all studied sections. The identification procedure is depicted for a specific section in Figure 2b), and the results of such a procedure are depicted for all sections in Figure 3.

The different sections were described in terms of lithological heterogeneity using the following geometrical

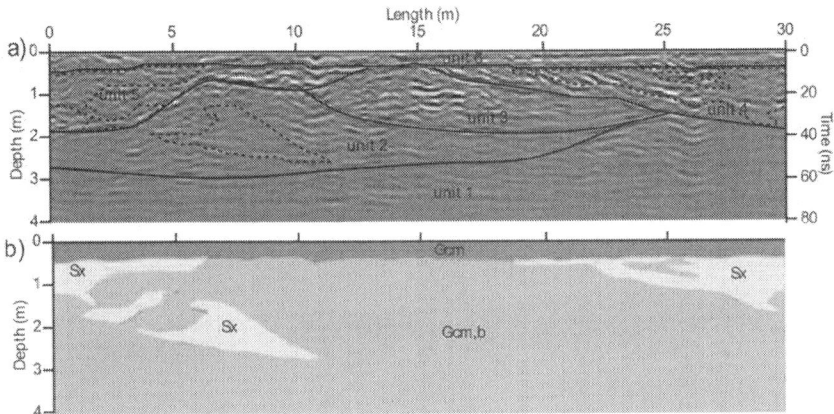

Figure 2. a) Representative GPR profile with antenna 200 MHz and major GPR reflection patterns; b) Interpretative sketches of major lithofacies (Sx: sand; Gcm: sandy gravel; Gcm, b: sandy gravel bimodal).

structural or architectural indicators. These were defined in agreement with the architecture of the studied material and sections which were split into three generic elements: the predominant material, referred to as "main lithofacies" which occupies most of the section, the upper layer and sand lenses (Figure 3). For these elements, the following indicators are proposed (Figure 3):

$$\begin{cases} F_M = \dfrac{S_M}{STC} \\ F_{La} = \dfrac{S_{La}}{STC} \\ F_{Le} = \dfrac{\sum_i A_i}{STC} \end{cases} \quad (4a)$$

$$IDC = \dfrac{\sum_i^n L_i}{L} \quad (4b)$$

$$IDCL = \dfrac{L_M}{L} \quad (4c)$$

$$L_M = \max(L_i) \quad (4d)$$

$$X_{GC} = \dfrac{\sum_{j=1}^n x_i A_i}{\sum_{j=1}^n A_i} \quad (4e)$$

where S_M, S_{La}, A_i and STC refer to the surface occupied by the main lithofacies, the upper layer, the ith sand lens and the total area of the whole section, respectively; L_i corresponds to the length of the ith sand lens and x_i to its abscissa (Figure 3). Indicators F_M, F_{La} and F_{Le} represent the degree of occupancy in the section by the main lithofacies, the upper layer (Figure 3, La) and sand lenses (Figure 3, Le), respectively; indicators IDC and IDCL quantify the degree of obstruction of the horizontal length by all inclusions or by the largest inclusion, and finally X_{GC} defines the gravity center of the sand lenses.

On average, the upper layer occupies ~17% of the whole section and sand lenses ~ 21%. For most sections, the degrees of occupancy were close to average, except Section S4 with low values for both upper layer and sand lenses, Section S1 with a low degree of occupation for sand lenses, and Section S5 with a high degree of occupation for both the sand and the

RESULTS AND DISCUSSION 103

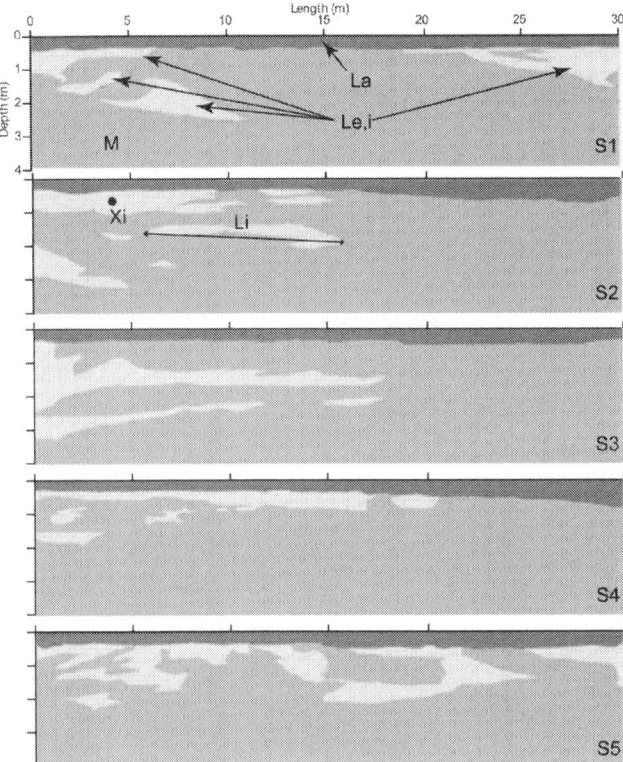

Figure 3. Schematic lithologic representation of the 5 sections under study: the sections are filed with the main lithofacies (M), a sandy gravel bimodal referred to as Gcm, b, an upper layer (La) made of sandy gravel referred to as Gcm, and sandy lenses (Le) made of sand referred to as Sx. The sandy lenses are characterized by their length (Li), their position (xi) and their area (Ai) for the lens i (i = 1 to n, with n the total number of lenses).

upper layer. For the position of lenses, the sections were ranked in function to the placement of sand lenses (Figure 3). For the first section, sand lenses are either close to the right or left boundaries, with a gravity center close to the center (i.e., X_{GC} = 13.2 » 15 m). Sections S2, S3 and S4 contain all sand lenses close to their left edge but with an increasing abscissa for the gravity center, corresponding to an increase in the degree of obstruction of the section length by sand lenses. The last section, S5, corresponds to a complete obstruction of section length, with the highest value of indicator IDC (Table 2). Clearly, sections exhibit different geometries and may influence flow in different ways.

Modeling water Redistribution during Drainage

We first modeled water redistribution in the whole profile during a drainage event. All soil profiles were considered close to saturation at the beginning of the event with a water pressure head of −0.01 m. Clearly, water redistribution depends on the profile. In Figure 4, the evolution of water content with time is illustrated for three contrasting profiles: a) uniform profile (Section H, Figure 4), b) heterogeneous profile with layered system

Table 2. values of geometrical parameters for the studied sections, H for homogeneous and the 5 heterogeneous sections from S1 to S5 are described in Figure 3.

Flow	F_{La}	F_{Le}	X_{GC} (m)	IDC	IDCL	L_{max} (m)
H	0	0	-	-	-	0
S1	12.4	12.4	13.20	1.01	0.37	11.24
S2	12.8	18.0	5.52	1.02	0.36	10.92
S3	12.2	20.7	6.35	1.12	0.60	17.91
S4	7.0	12.2	9.16	1.00	0.55	16.50
S5	16.6	21.2	13.22	1.22	0.50	13.99

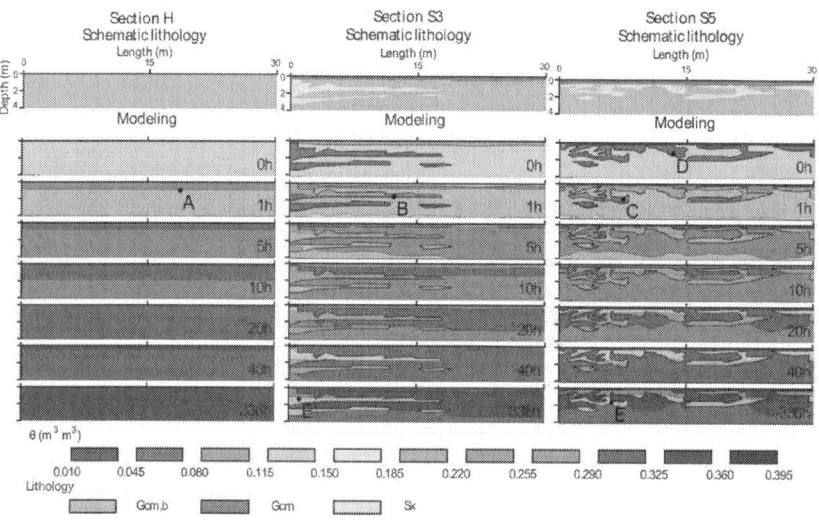

Figure 4. Time evolution of water content in three sections during the drainage phase; the sections and their respective geometry are reminded at the upper part of the figure.

and inclusions mainly on the left (Section S1, Figure 4) and finally a layered profile with inclusions randomly distributed along the whole section (Section S5, Figure 4). It is clear that in the uniform profile, isolines for water content are horizontal lines, revealing a piston type drainage with homogeneity of processes over the whole section (Figure 4, A). On the contrary, in heterogeneous sections, the geometry of isolines for water content is impacted by the geometry of inclusions and reveals that inclusions store water (Figure 4, B and Figure 4, C). At time zero, there is higher water content in sand lenses and in the upper layer (Figure 4, D). Indeed, at the beginning, all lithofacies are close to saturation and saturated water content is higher for the sand and the Gcm constituting the upper layer (Table 1, lithofacies S-x and Gcm, respectively). Saturated water content lies at 0.274 and 0.359, respectively for the upper layer and the sand lenses, versus 0.186 for the main lithofacies. Thus, it seems logical to get higher water content in the upper layer and the sand lenses at the beginning. For longer times, drainage remains the same in the main material, with water content similar to that obtained for the uniform section. On the other hand, drainage seems to slow down in the sand lenses, with higher water content during the whole drainage phase. When time equals 336 h, clearly the sand lenses store water in heterogeneous profiles S3 and S5 (Figure 4, E and Figure 4, E').

Water fluxes are now described and explained. At upper boundary, water fluxes are null by definition (drainage and no flux entering at surface). At lower boundary, fluxes decrease with time and tend towards zero (Figure 5b)). The differences between sections are really tiny. At the beginning of the process (i.e., between 0 and 5 h), flow rate is a little bit faster in the homogeneous section (Figure 5b)). After ten hours, the flow rate for the homogeneous sections decreases faster and becomes slightly below flow rates for the other sections. The following conclusions can be stated on the basis of the analysis of cumulative fluxes (Figure 5a)). Less amounts of water are drained and drainage is a little bit faster for the homogenous section. The comparison of different heterogeneous sections shows that there is a variability in drained cumulative fluxes (Figure 5a) and Figure 5b)). Yet, despite these slight differences, there is not a great impact from the section's heterogeneity on flux rate at the lower boundary conditions, meaning that fluxes during the drainage and water redistribution phases should not be greatly impacted by the heterogeneity of the deposit.

Finally, at surface and at the lower boundary, water pressure heads decrease for all sections, as the result of water drainage and redistribution in the profile. This decrease is slightly more pronounced for the case of the homogenous section. For heterogeneous section, water pressure head at surface decreases to a lesser extent; revealing an over-pressure at surface, in comparison to the case of the homogeneous section. This over-pressure witnesses the increase in flow impedance for the heterogeneous section.

This is logical and in agreement with the fact that heterogeneous profiles are made of the main material plus two materials with lower saturated hydraulic conductivity [25] [26].

Modeling water Infiltration during Rainfall

Water infiltration was modeled for a particular rainfall event, considering atmospheric condition applied at surface boundary with a flux of 0.3438 m/h, which corresponds to a rainfall intensity of 6 mm/h, relating to a period of intense rain. The analysis of water content profiles gives information on the flow pathways during infiltration. For the homogeneous profile, wetting fronts correspond to horizontal lines, revealing a piston type infiltration (data not shown). Water infiltrates progressively with similar fluxes at a fixed depth. In the case of heterogeneous sections, the wetting fronts are no longer horizontal. During infiltration, for times between 0 and ~3 h, the wetting fronts strongly depend on the geometry of inclusions (Figure 6). For Section S1, water infiltrates in the medium of the section between the inclusions (Figure 6, A). For Section S3, water infiltrates mainly on the

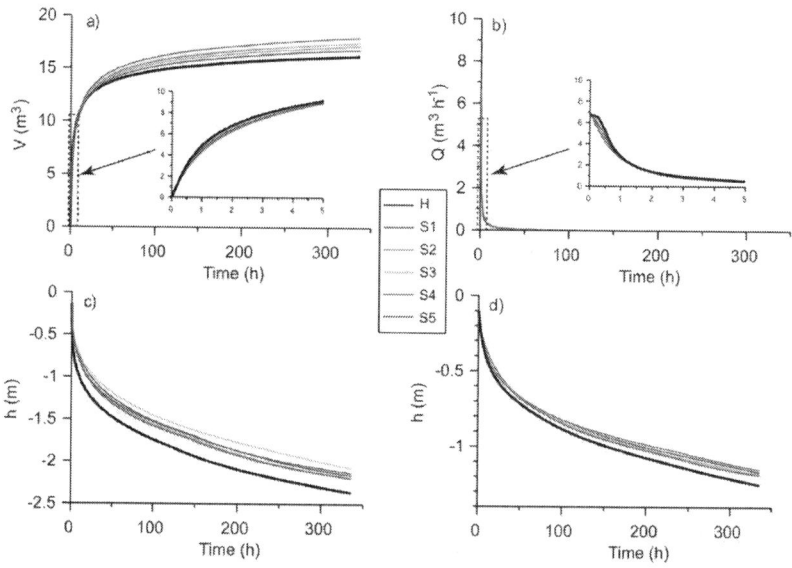

Figure 5. Water redistribution during drainage phase: a) total cumulative flux drained at lower boundary, i.e., at a depth of 4 m, b) water flux at the lower boundary, water pressure heads at surface c) and at lower boundary d).

RESULTS AND DISCUSSION 107

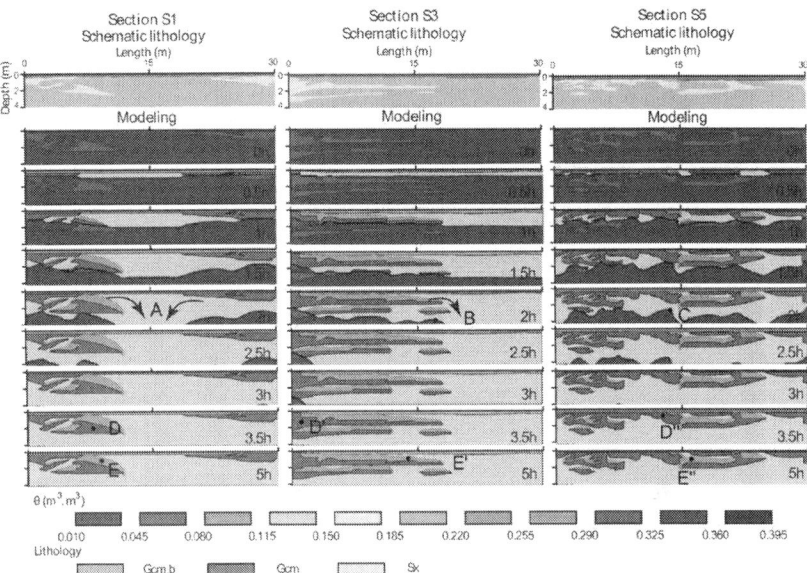

Figure 6. Time evolution of water contents in three sections during the studied rainfall event.

right side, apart from sand inclusions (Figure 6, B). And finally, for the more complicated Section S5, there is no specific preferential flow since inclusions are spread over the whole length of the profile (Figure 6, C). But, locally, water avoids inclusions. At the end of infiltration, when the wetting fronts have reached the lower boundary, water content differs between the main lithofacies and the inclusions. The sand lenses exhibit higher water content (Figure 6, D, D' and D") because of their high saturated water content (Table 1). Above the edges of sand lenses, water content is increased (Figure 6, E, E' and E"). This results from the local increase in pressure head at the interfaces between sand lenses and the main material, above sand lenses. Such an increase reveals a drop in hydraulic conductivity from the main material to the sand. In fact, under these conditions, the sand lenses exhibit a lower hydraulic conductivity and force water to avoid them and triggers the increase in water pressure head. This increase above sand lenses can be regarded as similar in process to the formation of a perched aquifer [26]. Such results are in agreement with the previous works [3]. The contribution of this study is to show that depending on the section considered and the related sedimentological heterogeneity, flow pathways may change.

Soil heterogeneity affects flow pathways but also water infiltration at surface. For water pressure head, the influence is quite small (Figure 7c)).

All water pressure heads tend towards zero and the difference between the homogeneous and the heterogeneous sections is quite tiny. For fluxes, the heterogeneity impedes flow rate and reduces water infiltration. In the homogeneous section, the whole water flux applied at surface is infiltrated and the water flux entering the profile equals the flux imposed (Figure 7a)). For the other sections, the water flux imposed at surface is completely infiltrated during one hour before being reduced, leading to a decrease in infiltration rate as depicted in Figure 7a). Sorting sections as a function of infiltrated fluxes leads to: H > S3 > S2 > S1 > S4 » S5. Such ranking will be discussed below in light of the following results.

At a depth of 4 m (i.e., lower boundary), both water fluxes and water pressure heads are impacted. For the homogeneous section, water pressure head and fluxes suddenly increase around 2 h. This time corresponds to the time needed for the wetting front to reach the lower boundary. It can be stated that, for the homogeneous section, the wetting front is sharp as witnessed by the sudden and drastic increase in both water pressure and

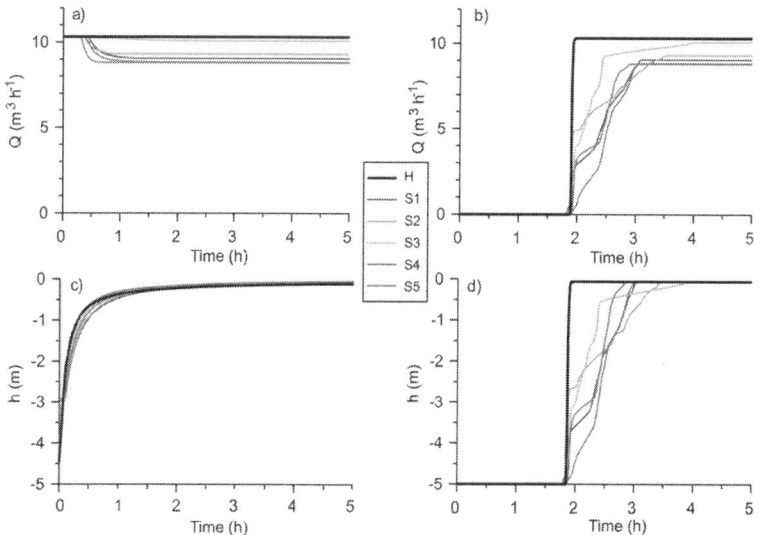

Figure 7. Water infiltration during rainfall event: fluxes at surface a) and at the lower boundary, i.e., at 4 m depth b); water pressure heads at surface c) and at lower boundary d).

flux rates. At the end, the whole profile is saturated with pressure heads close to zero and the water infiltrates at the same flow rate along the whole profile (gravity driven saturated flow). The heterogeneity of the section impacts the arrival of the wetting front. These arrive slightly later, around

RESULTS AND DISCUSSION

2.25/2.5 h and the increase of water pressure head and water flux rate is less sharp with more complex shapes (Figure 7b) and Figure 7d)). In fact, these more complex shapes reveal several arrivals corresponding to several different pathways. In this case, the complexity of flow pathways due to profile heterogeneity impacts significantly the fluxes, including the fluxes that occur at every depth in the profile. Once the wetting front has reached the lower boundary, water pressure head increases up to zero as for the upper boundary, leading to full saturation along the whole profile. The water flux at the lower boundary tends towards the actual infiltration rate at the upper boundary. For instance, the water flux at the lower boundary of Section S4 is around 8.8 $m^3 h^{-1}$, corresponding to the water flux that infiltrate sat surface for an applied water flux of 10.3 $m^3 h^{-1}$. Regarding the sorting of arrival times, we obtain respectively the following rankings: S3 < S2 < S1 = S4 < S5. We obtain a ranking similar to that of infiltration rates, meaning that the section in which water infiltrates the less at surface corresponds to the slowest velocity of water and wetting fronts in the profile. Regarding time dispersion around averaged arrival times, the ranking changes without any link between arrival time and time dispersion. Indeed, some sections exhibit short arrival time with low dispersion (Section S2), others exhibit longer arrival time with large dispersion, etc.

Briefly, the heterogeneity of soil profile impacts significantly flow pathways and water infiltration and runoff at surface. In the studied profiles, the upper layer and the inclusions impede flow, resulting in more runoff at surface and slower movement of wetting fronts. In addition, these trigger the establishment of several flow pathways, resulting in the succession of several arrivals for the wetting front at a fixed depth. These results show that the heterogeneity of soil may significantly impact water redistribution or infiltration during both drainage and rainfall events. The following section addresses the effect of soil heterogeneity for a complete meteorological chronic.

Modeling water Infiltration for Meteorological data

A complete meteorological data was modeled for the several sections and for the first 1000 h of year 2008. Figure 8 presents water pressure head and cumulative flux at surface and at the lower boundary (Figures 8a)-d)). Figure 8 depicts also the water fluxes at boundaries for a specific rainfall event (Figures 8e)-f)). The meteorological

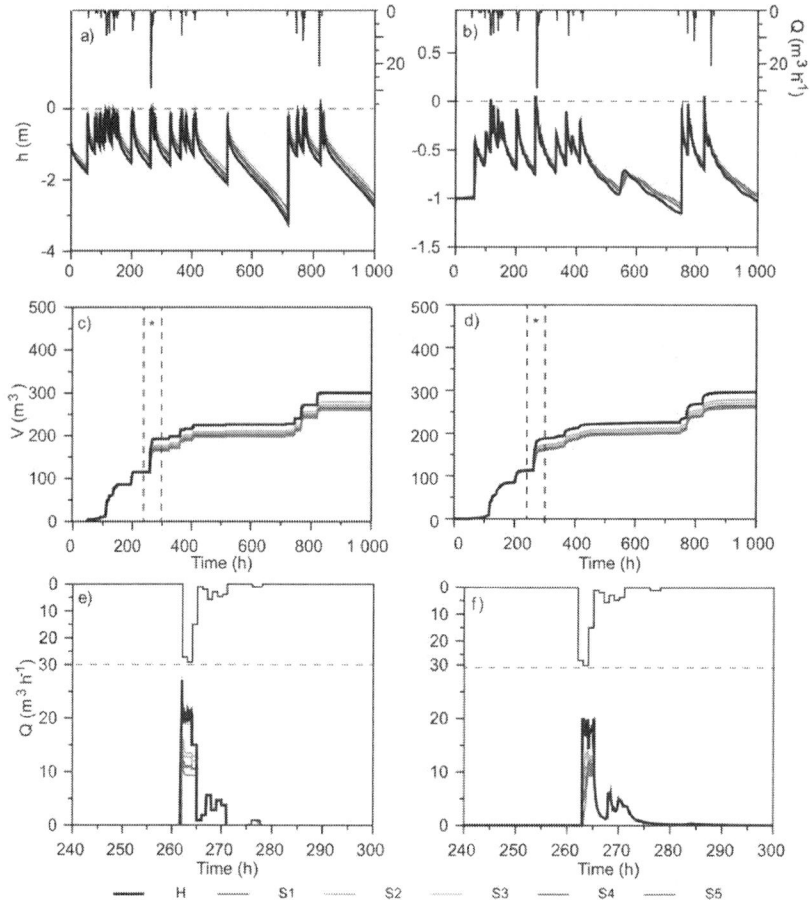

Figure 8. Numerical results for the meteorological chronic, potential flux at surface (a) or b)), water pressure heads at surface a) and at lower boundary, i.e., at a depth of 4m b), real cumulative fluxes at surface c) and at lower boundary d), zoom on a rainfall event depicting flux at surface e) and at lower boundary f).

data corresponds to the succession of rainfall events and drainage phases (dry periods) in between, measured at the meteorological station of Bron in Lyon urban areas.

For all sections, rainfall events logically trigger a quick increase in water pressure head and drainage triggers a slower decrease in water pressure head. It can be noted that time evolution of pressure head is wider at surface with values between −3 m and 0 m (Figure 8a)), whereas at the lower boundary the water pressure head varies between −1.5 and zero (Figure 8b)). This is often observed in most cases [27]. The soil can be seen as a buffer and variations in hydric conditions are often wider at surface than in deeper horizons. The differences between the homogeneous section and the heterogeneous sections are in agreement with what was described above for case studies. The heterogeneity impacts surface water pressure heads more during drainage phases than during rainfall events (Figure 8a)), as already stated above. The same trends are observed for water pressure head at lower boundary level.

For water fluxes, only cumulative fluxes were reported in Figure 8c) and Figure 8d) for the sake of clarity. Every rainfall event logically triggers a sharp increase in cumulative infiltration, whereas the drainage corresponds to plateaus. It can be noted that the shapes are sharper for cumulative infiltrations at surface (Figure 8c)) in comparison with the lower boundary (Figure 8d)). This results from the buffer effect of the soil and the fact that processes responsible for water movement in the soil smoothen the variation of fluxes and water pressure heads in deeper horizons. The difference between all the sections is extremely tiny before 250 min. After, the sections split into several groups with a clear distinction between the homogeneous and the heterogeneous sections. As observed in the previous chapter, the homogeneous section infiltrates more water than the others. This becomes clear when analyzing one specific event (Figure 8e) and Figure 8f)). For this event occurring between 260 min and 270 min, water infiltration at surface is the most important in the homogeneous section. For all the heterogeneous sections, water infiltration decreases, and sections can be sorted as a function of infiltrated fluxes as follows: H > S3> S2 > S1 » S4 > S5, which corresponds to the ranking obtained above for case study of rainfall event. Clearly, the behavior of each section in terms of water pressure head and fluxes corresponds to the same behavior as described above for a particular drainage and rainfall event, and the same conclusions can be stated.

At the end of the chronic, the combination of similar effects for all events leads to significant effect on the water budget. For instance, profile heterogeneity impacts the percentage of cumulative fluxes that was infiltrated at surface and that achieved infiltration to the lower boundary (positioned at a depth of 4 m). For the case of the homogeneous section, 95.2% of the water infiltrates into the profile with only 4.6% as runoff. The heterogeneity simultaneously increases runoff and decreases water infiltration. The ranking of sections leads to: H < S3 < S2 < S1 < S4 < S5 with a maximum value for runoff at 17.7% and a minimum value of runoff at 11.6%, for heterogeneous sections. Such a ranking is in full agreement with the previous analysis of case studies. Clearly, soil heterogeneity impacts runoff at surface but it can also be concluded also that depending upon the location of sections in the field and the geometry of inclusions, runoff may change significantly, as already proved [25] [26] even for this kind of deposit [3] [4]. But this study clearly shows that for the same deposit, soil heterogeneity may vary and flow pattern may vary as well. It is then crucial to account for spatial variability of the profile heterogeneity at the basin scale.

Link between flow and Geometrical indicators of heterogeneous Sections

In the previous sections, it was clearly demonstrated that the comparison of sections gave the same results for all drainage and rainfall events and over a chronic of 1000 h. It was shown that heterogeneity affected flow in the same way, but that the degree of heterogeneity of the profile plays on the intensity of the effect of soil heterogeneity. As a consequence, characterizing sedimentological properties for only one section and modeling unsaturated flow for a sole section is not sufficient for flow rate prediction at the infiltration basin scale. In this section, we assess the link between the impact of heterogeneity on flow and the values of indicators. As the influence of profile heterogeneity on flow pattern remains the same, whatever the drainage or rainfall event, we need only to study the link between flow and indicators for one case. Indeed, similar trends will be found for any other rainfall event or even for the meteorological chronic.

In this part, we assess the link between geometrical indicators and runoff at surface obtained for the complete meteorological chronic. In terms of runoff at surface, we obtain the following ranking: H < S3 < S2 < S1 < S4 < S5. Here, we deal with the effect of soil heterogeneity on flow impedance. At first, we test the indicator F_{La}, which corresponds approximately to the thickness of the upper layer. Indeed, physically, if we consider that the lower saturated hydraulic conductivity of the upper layer is one of the main key factors for flow impedance, we should get a clear link between the ranking of sections with regards to runoff and to F_{La}. Yet, these rankings do not correspond. The ranking with regards to F_{La} gives: H < S4 < S3 < S1 < S2 < S5; which is quite different. As a second step, we consider that the sand lenses are one of the main key factors for flow impedance. We sorted the sections as a function of the degree of occupation of the section by the sand lenses, i.e., indicator F_{Le}. The ranking obtained with F_{Le} was quite different to that of the runoff (Table 3). The procedure was repeated to all indicators and no clear concordance was found (Table 3). Briefly, there is no link between runoff and one particular indicator. This shows that flow may be more the result of a combination of several indicators. Clearly, more data for more sections are needed to identify the combination of proposed parameters that explain results related to flow. Regarding the physics of flow, the worse situation for water infiltration corresponds to the section for which the upper layer is thick and sand lenses occupy a large fraction of the section and are spread all over the section length. For instance, Section S5, which fulfills this criterion, triggers the smallest runoff and corresponds to the strongest effect of heterogeneity on flow.

Table 3. Ranking of section with regards to runoff for the meteorological data (Flow) and with regards proposed geometrical indicators (F_{La}, F_{Le}, X_{GC}, IDC, IDCL, and L_{max}).

Flow	F_{La}	F_{Le}	X_{GC} (m)	IDC	IDCL	L_{max} (m)
S						
H	H	H	-	-	-	H
S3	S4	S4	S2	S4	S2	S2
S2	S3	S1	S3	S1	S1	S1
S1	S1	S2	S4	S2	S5	S5
S4	S2	S3	S1	S3	S4	S4
S5	S5	S5	S5	S5	S3	S3

CONCLUSION

In this study, we examined the effect of soil heterogeneity on unsaturated flow processes. Numerical modeling was employed to model unsaturated flow for different events (drainage phase, rainfall event and a succession of rainfall and drainage events for a complete meteorological chronic) in 5 different real sections and a hypothetic homogeneous section filled with the main lithofacies (blank). These sections were previously properly character- rized with regard to their sedimentological properties (architecture and lithofacies) using GPR and then modeled in terms of flow. It was proved that the upper layer and sand inclusions impeded flow at the section scale, reducing water movement into the soils and increasing water runoff at surface. Locally, the inclusion of sand lenses modified flow pathways, forcing the water to pass through pathways among inclusions. These effects were due to their lower hydraulic conductivity. In addition, the upper layer and sand lenses increased accumulation of water in the profile due to their higher saturated water content. One of the main conclusions of this study is that if all heterogeneous sections differ from those of the homogeneous section, the variability among heterogeneous sections is also significant. Clearly, a sole section is not representative of all sections and several sections must be investigated if flow needs to be investigated at the scale of the whole infiltration basin. Thus, we tried to propose geometrical indicators to explain and potentially predict flow pattern. However, the results show that the link between these indicators and modeled flow is not direct and that a combination of indicators rather than a sole indicator should be identified for the prediction of flow. Additional data are required to properly define and identify the right combination of factors, which is the subject of ongoing research. Finally, this paper clearly demonstrates the influence of soil heterogeneity on flow pathways, which is essential for understanding pollutant transport by water and access to the reactive particles of soils [9] [10] . This gives pertinent information for the modeling and understanding of the fate of pollutants in such a deposit and in infiltration basins that rest on this kind of deposit.

ACKNOWLEDGEMENTS

The first author received a Ph.D. scholarship from the projects CAPES/COFECUB and FACEPE. The authors are grateful to the OTHU, Greater Lyon and the ANR-GESSOL program (FAFF project: Filtration Function of an Urban Structure—Consequence on the Formation of an Anthroposol) for their logistic and financial support.

REFERENCES

1. Mikkelsen, P.S., Hafliger, M., Mikkelsen, P.S., Hafliger, M., Ochs, M., Jacobsen, P., Tjell, J. C. and Boller, M. (1997) Pollution of Soil and Groundwater from Infiltration of Highly Contaminated Stormwater—A Case Study. Water Science and Technology, 36, 325-330.
2. Goutaland, D. (2008) Caracterisation hydrogeophysique d'un depot fluvioglaciaire. Evaluation de l'effet de l'heterogeneite hydrodynamique sur les ecoulements en zone non-saturee. Ph.D. Thesis, Graduate School of Chemistry, University of Lyon, 241 p.
3. Goutaland, D., Winiarski, T., Lassabatere, L., Dube, J.S. and Angulo-Jaramillo, R. (2013) Sedimentary and Hydraulic Characterization of a Heterogeneous Glaciofluvial Deposit: Application to the Modeling of Unsaturated Flow. Engineering Geology, 166, 127-139.
4. Winiarski, T., Lassabatere, L., Angulo-Jaramillo, R. and Goutaland, D. (2013) Characterization of the Heterogeneous Flow and Pollutant Transfer in the Unsaturated Zone in the Fluvio-glacial Deposit. Procedia Environmental Sciences, 19, 955-964.

REFERENCES

5. Leclerc, R.J. and Hickm, E.J. (1997) The Internal Structure of Scrolled Floodplain Deposits Based on Ground-Penetrating Radar, North Thompson River, British Columbia. Geomorphology, 21, 17-38.
6. Mumphy, A.J., Jol, H.M., Kean, W.F. and Isbell, J.L. (2007) Architecture and Sedimentology of an Active Braid Bar in the Wisconsin River Based on 3-D Ground Penetrating Radar. Special Paper of the Geological Society of America, 432, 111-131.
7. Lunt, I.A., Bridge, J.S. and Tye, R.S. (2004) A Quantitative Three-Dimensional Depositional Model of Gravelly Braided Rivers. Sedimentology, 51, 377-414.
8. Lassabatere, L., Angulo-Jaramillo, R., Soria Ugalde, J.M., Cuenca, R., Braud, I. and Haverkamp, R. (2006) Beerkan Estimation of Soil Transfer Parameters through Infiltration Experiments—BEST. Soil Science Society of America Journal, 70, 521-532.
9. Lassabatere, L., Winiarski, T. and Galvez Cloutier, R. (2004) Retention of Three Heavy Metals (Zn, Pb, and Cd) in a Calcareous Soil Controlled by the Modification of Flow with Geotextiles. Environmental Science and Technology, 38, 4215-4221.
10. Lamy, E., Lassabatere, L., Bechet, B. and Andrieu, H. (2009) Modeling the Influence of an Artificial Macropore in Sandy Columns on Flow and Solute Transfer. Journal of Hydrology, 376, 392-402.
11. Kohne, J.M., Kohne, S. and Simunek, J. (2009) A Review of Model Applications for Structured Soils: b) Pesticide Transport. Journal of Contaminant Hydrology, 104, 36-60.
12. BURGEAP (1995) Study of the Water Table of the Eastern Part of Lyon Area. (In French) Le Grand Lyon Direction de l'eau, Lyon.
13. Huggenberger, P., Meier, E. and Pugin, A. (1994) Ground-Probing Radar as a Tool for Heterogeneity Estimation in Gravel Deposits: Advances in Data-Processing and Facies Analysis. Journal of Applied Geophysics, 31, 171-184.
14. Beres, M., Huggenberger, P., Green, A.G. and Horstmeyer, H. (1999) Using Two- and Three-Dimensional Georadar Methods to Characterize Glaciofluvial Architecture. Sedimentary Geology, 129, 1-24.
15. Bridge, J.S. and Lunt, I.A. (2006) Depositional Models of Braided Rivers. In: Sambrook-Smith, G.H., Best, J.L., Bristow, C.S. and Petts, G.E., Eds., Braided Rivers: Process, Deposits, Ecology and Management, Blackwell Publishing, Oxford, 11-49.
16. Schrott, L. and Sass, O. (2008) Application of Field Geophysics in Geomorphology: Advances and Limitations Exemplified by Case Studies. Geomorphology, 93, 55-73.
17. Miall, A.D. (1999) Principles of Sedimentary Basin Analysis. 3rd Updated and Enlarged Edition, Springer, Berlin.

18. Goutaland, D., Winiarski, T., Dube, J.S., Bievre, G., Buoncristiani, J.F., Chouteau, M. and Giroux, B. (2008) Hydrostratigraphic Characterization of Glaciofluvial Deposits Underlying an Infiltration Basin Using Ground Penetrating Radar. Vadose Zone Journal, 7, 194-207.
19. Goutaland, D., Winiarski, T., Bièvre, G. and Buoncristiani, J-F. (2005) Interêt de l'approche sedimentologique en matière d'infiltration d'eaux pluviales. Caracterisation par radar geologique. Techniques Sciences Methodes, 10, 71-79.
20. Mitchum, R.M.J., Vail, P.R. and Sangree, J.B. (1977) Stratigraphic Interpretation of Seismic Reflection Patterns in Depositional Sequences. In: Payton, C.E., Ed., Seismic Stratigraphy: Applications to Hydrocarbon Exploration, the American Association of Petroleum Geologist, Tulsa, Vol. 26, 117-133.
21. Simunek, J. and van Genuchten, M.T. (2008) Modeling Nonequilibrium Flow and Transport Processes Using Hydrus Version 4.0. HYDRUS 1D. Department of Environmental Sciences, University of California, Riverside.
22. van Genuchten, M.T. (1980) A Closed-Form Equation for Predicting the Hydraulic Conductivity of Unsaturated Soils. Soil Science Society of America Journal, 44, 892-898.
23. Lassabatere, L., Angulo-Jaramillo, R., Goutaland, D., Letellier, L., Gaudet, J-P., Winiarski, T. and Delolme, C. (2010) Effect of the Settlement of Sediments on Water Infiltration in Two Urban Infiltration Basins. Geoderma, 156, 316-325.
24. Heinz, J., Kleineidam, S., Teutsch, G. and Aigner, T. (2003) Heterogeneity Patterns of Quaternary Glaciofluvial Gravel Bodies (SW-Germany): Application to Hydrogeology. Sedimentary Geology, 158, 1-23.
25. Birkholzer, J. and Tsang, C.F. (1997) Solute Channeling in Unsaturated Heterogeneous Porous Media. Water Resources Research, 33, 2221-2238.
26. Miyazaki, T. (2006) Water Flow in Soils. CRC Press, Boca Raton.
27. Kutilek, M. and Nielsen, D.R. (1994) Soil Hydrology. Catena Verlag, Cremlingen.

CITATION

Artur PaivaCoutinho,LaurentLassabatere,ThierryWiniarski,Jaime Joaquim da Silva PereiraCabral,Antonio Celso DantasAntonino,RafaelAngulo-Jaramillo, (2015) Vadose Zone Heterogeneity Effect on Unsaturated Water Flow Modeling at Meso-Scale. *Journal of Water Resource and Protection*, **07**, 353-368. doi: 10.4236/jwarp.2015.74028

CHAPTER 5

Hydrodynamic Performances of Air-Water Flows in Gullies with and Without Swirl Generation Vanes for Drainage Systems of Buildings

Der-Chang Lo [1], Jin-Shuen Liou [1] and Shyy Woei Chang [2,*]

[1]Department of Maritime Information and Technology, National Kaohsiung Marine University, No. 142, Hai-Chuan Road, Nan-Tzu District, Kaohsiung 811, Taiwan; E-Mails: loder@mail.nkmu.edu.tw(D.-C.L.); jasonpoter@gmail.com (J.-S.L.)
[2]Thermal Fluids Laboratory, National Kaohsiung Marine University, No. 142, Hai-Chuan Road, Nan-Tzu District, Kaohsiung 811, Taiwan

ABSTRACT

As an attempt to improve the performances of multi-entry gullies with applications to drainage system of a building, the hydrodynamic characteristics of air-water flows through the gullies with and without swirl generation vanes (SGV) are experimentally and numerically examined. With the aid of present Charge Coupled Device (CCD) image and optical systems for experimental study, the mechanism of air entrainment by vortex, the temporal variations of airflow pressure, the trajectories of drifting air bubbles and the self-depuration process for the gullies with and without SGV are disclosed. The numerical simulations adopt Flow-3D commercial code to attack the unsteady two-phase bubbly flows for resolving the transient fields of fluid velocity, vorticity and pressure in the gullies with and without SGV. In the twin-entry gully without SGV, air bubbles entrained by the entry vortex interact chaotically in the agitating bubbly flow region. With SGV to trip near-wall flows that stratify the drifting trajectories of the air bubbles, the air-bubble interactions are stabilized with the discharge rate increasing more than 7%. The reduction of the self-depuration period by increasing discharge rate is observed for the test gullies without and with SGV. Based on the experimental and numerical results, the characteristic hydrodynamic properties of the air-water flows through the test gullies with and without SGV are disclosed to assist the design applications of a modern drainage system in a building.

INTRODUCTION

To facilitate the efficient water supply and discharge for a building remains as a difficult task due to the complex flow bifurcations in water supply networks as well as the dynamic and unsteady interfacial air-water flow mechanisms developed in a drainage system. For preventing odor transmissions into habitat spaces through a drainage network, the gullies that reserve a water seal for many discharge branches have demonstrated their convenience for installation and maintenance, with opportunities to simplify the drainage system. A recent growing rate for the usage of gullies in Taiwan has proven their potential benefits for building industries. For each device installed in a drainage system, its impacts on the system stabilities, in particular on the variations of airflow pressures responsive to the intermittent discharge(s) through a drainage piping system, have to be identified prior to its widespread applications.

Unlike a siphonic roof drainage system, the random and intermittent falling water into the vertical stack via the various discharge branches in a drainage system is not generally at the full water condition but entrains airflow to formulate a variety of complex air-water flows with various two-phase flow patterns. The interfacial air-water flow structures are affected by the geometries of the pipe-line and appliances, the flow rate and the location in a drainage system. In a branch and the vertical stack of a building drainage network, the interfacial flow structures are typical of intermittent stratified, wavy and annular flows [1]. The momentum changes of air-water flows caused by varying flow direction, expansions and contractions, bifurcations and/or chocking the airways incur the locally positive or negative transient airflow pressures that propagate throughout the entire drainage system at the sonic speed [2]. The impacts of such transient propagation—including the effects on acoustic resonances, discharging capacities and local negative or positive pressures—depend on the air-water interfacial structures and on the reflection and transmission of pressure waves on the interfacial and solid boundaries. Following a transient water discharge from the branch into the vertical stack of a drainage system, the considerable pressure oscillations at the elbow bend of the vertical stack were demonstrated to affect the entire drainage network [2]. At locations where the water curtain or excursion develops to intermittently block a high momentum air stream, the trap seal is often diminished by the raised positive airflow pressure due to the water hammer effect [3]. As the water seal prevents the transmission of foul odors ingress into the habitable spaces through the interconnected drainage network in a building, the survival of each water seal during random discharges is of primary importance. The various design codes for

INTRODUCTION

architectures normally request a trap seal with about 50 mm water height corresponding to the permissible pressure excursion of $\pm$ 375 N·m^{-2} [1,2]. To achieve this design goal, the relevant experimental and numerical works have being carried out. As an attempt to suppress the positive pressure surges in a drainage system, the propagations of air pressure transient in a simulated drainage system by solving the St. Venant equations using the finite difference scheme was numerically performed [4]. With the complex two-phase air-water flow structures in a drainage system, the suppression of undesirable pressure transients still remains as a formidable task. In particular, the air-water flow phenomena in the various types of components and appliances of a drainage system are interdependent, leading to complicated interactive hydrodynamic responses [1,2,3,4]. As an attempt to moderate the positive airflow pressure surges initiated from the bottom elbow bend of a vertical stack [3,4], the pressure accumulator was installed to provide additional expansion space for alleviating the positive airflow transients [4]. The streamlined vortex fin(s) with sidewall grooves [3] was installed at the elbow-bend of a vertical stack to induce longitudinal swirls for penetrating the downstream water curtain developed in the elbow-bend. With the numerical schemes for attacking the two-phase flow problems in a drainage system [5,6,7,8], the entrainment model was developed for solving the hydrodynamic characteristics of multi-phase flows involving hydraulic jumps with air entrainments [7,8]. With the presence of entrained air to add the damping effect on the collapsing bubbles, the damages caused by cavitation were alleviated, thus recommending the installation of aeration devices to entrain air for alleviating the cavitation effect.

In view of a gully within which the common water seal for many discharge branches is trapped, the hydrodynamic characteristics for the through air-water flow are further complicated and dependent on the geometries of the flow pathways. In [9,10], the experimental measurements for the flow dynamics and the numerical simulations for the dynamic responses in the multi-outlet siphonic roof drainage systems were respectively reported. The fundamental air-water flow phenomena in the multi-entry gully were illustrated using a set of numerical results simulated by Flow 3-D code [11]. Based on the assumption of lumped bubbly flow for the multi-entry gully, the geometries of entry and discharge ports as well as the plenum chamber were shown as the predominant factors to affect the hydrodynamic performances for this type of multi-entry gullies [11]. Driven by the need to miniaturize the multi-entry gully for building applications, a streamlined bump [12] was fitted at the location downstream the discharge port. With the locally siphonic effects at the throat of the partitioned discharge port, the upstream air-water flows were substantially stabilized; while the maximum flow rates were limited by the choking nozzle effect at the discharge port. In order to raise the maximum discharge capacity for the shallow type multi-entry gully, a ring of SGV (swirl generation vanes) is fitted in the annular flow pathway for stabilizing the air-water flows by stratifying the air-bubble drifting trajectories along the swirl induced by the SGV. This study adopts experimental and numerical methods to probe into the air-water flow phenomena taking place in the shallow-type twin-entry gullies without and with SGV. The flow phenomena, in particular for the dynamic air-water interfacial flow structures, disclosed by this work are beneficial for gully design practice with the follow-on researches directing toward the acoustic aspect of flow induced vibrations and the miniaturization of gully with optimized discharge rate. In what follows, the experimental and numerical methods are briefly illustrated and followed by a set of selective results to comparatively examine the SGV effects on the hydrodynamic performances for this type of gully.

RESEARCH METHODS

Experimental Apparatus and Test Details

Figure 1 depicts (a) test facilities with the optical device measuring the self-depuration performance (b) a twin-entry test gully with SGV. As depicted by Figure 1a, the supplied water from tank (1) is located at second floor of the in-house fifth-floor height drainage test facility, giving rise to the pressure potential of 1.2 m of water height to facilitate the required flow rates for experimental tests. As indicated in Figure 1a, the fresh water fed from tank (1) flows through a vertical stack (2) to the twin-entry test gully (3) via two horizontal entry pipes tangent to the gully drum. The present drainage system is complied with the new construction method using the single-pipe vertical stack with the Air Admitting Valve (AAV) (4) installed on top of the vertical stack. Airflow pressures are controlled in the typical range of ± 375 Nm-2 via the auto air entrainments through the AAV (4) shown in Figure 1a. The net volume of water flow through the test gully (3) in Figure 1a is measured by the downstream water tank (5) with the time span detected by the electronic timer for accounting the averaged water flow rate through the test gully (3). A scale attached along the inner periphery of each transparent inlet pipe (6), (7), as indicated in Figure 1a, detects the water flow level for the stratified entry air-water flow in the horizontal branches (6), (7). The void fraction (α) of each entry mixed water stream can be accordingly determined. The air-water flow structures in the gully at each tested water flow rate at single- and/or twin-entry flow conditions are visualized from the snapshots imaged by the Charge Coupled Device (CCD) system. This imaging system records the flow snapshots at 300 fps with 600 pixels per gully width. The CCD camera (8) shown in Figure 1a is aimed at the angle normal to the test gully (3) with a constant focal length. The static airflow pressure is detected by a computerized digital micro manometer (9) in Figure 1a with the precision of 0.01 mm H_2O. As indicated in Figure 1b, the pressure tap measuring the airflow pressure above the entry vortex of the test gully is located on the frame attached on the top plane of the test gully with the probing depth to be precisely measured. Another port of the digital micro manometer is vented to atmosphere so that the static airflow pressures at the measuring locations above the entry vortex are detected. This type of pressure measurement device utilizes the piezoelectricity to convert pressure signal into electrical potential. The pressure measurements are synchronously recorded with the flow images taken by the CCD system, which are constantly monitored by the on-line data acquisition system. The detailed temporal variations of the airflow pressure and the corresponding flow images detected at each test condition are simultaneously recorded for post data processing. The test gully is made from a transparent arctic block. At each pre-defined flow test condition, a light sheet is emitted toward the dyed test gully behind which the photometric receiver is installed to detect and record the temporal lumen variations. By way of analyzing the temporal photometric variation, which is responsive to the temporal variation of dye concentration within the test gully, the self-depuration performance is revealed.

Figure 1b depicts the twin-entry test gully with SGV. As shown in Figure 1b, the test gully is configured by a vertical primary drum that directs the entry mixed water streams from the horizontal twin-entry ports in the downward direction toward the gully base. The radial spreading air-water stream then sharply turns and flows upward in the annular pathway between the primary and secondary drums. Over the circumferential band on the outer cylindrical wall of the secondary drum, ten SGV are in-line arranged and oriented at 45 deg. relative to the upward stream. These vanes are fitted to trip the anti-clockwise annular swirl between the primary and secondary drums. The cross-section area of discharge port is equal to the sectional annular area between the primary and secondary drums. The upward air-water stream is spilled out of the annular pathway toward the discharge port. As the overlapping height between the primary and secondary drums is 50 mm, the minimum water seal height in the test gully is ensured above than 50 mm. A replaceable filter leaf is installed above the cylindrical core on the top of the test gully, which permits the air entrainments from the surrounding atmosphere. As the two entry ports are in tangent with the outer rim of the gully casing, a central vortex is induced in the primary drum after feeding the mixed water flow into the gully. The free surface of the entry vortex formulates the airway to entrain air into the liquid pool, which will be later demonstrated. It is noticed that present orientation for the SGV is attempted to induce the co-current swirl at the same direction as the free vortex formulated in the primary drum. With the co-current swirl in the annular flow pathway in which the air-water stream flows in upward direction, the drifting air bubbles are guided by the near-wall flows over the roughened cylindrical wall on the secondary drum.

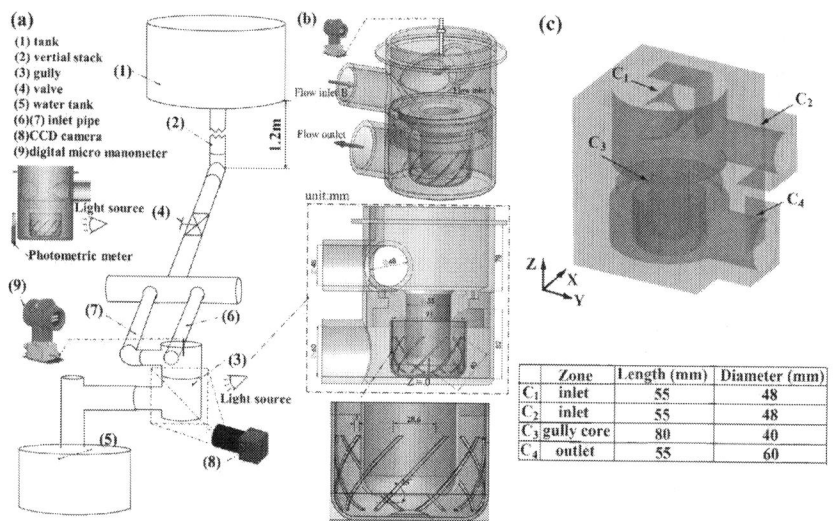

Figure 1. (a) test facilities; (b) twin-entry test gully with swirl generation vane (SGV); (c) layout of numerical model.

In order to examine the self-depuration performance of a gully, a set of optical device [12] is adopted to detect the temporal variations of the lumen level shaded by the dyed test gully. In the attempt to measure the self-depuration performance for the bulk flow of a test gully, the relative self-depuration properties are comparatively evaluated by measuring the temporal L/L_1 variations. The photometric meter adopted by this work is a two-dimensional device, which is attached on the transparent cylindrical casing of the test gully as indicated in Figure 1a. As the air entrained into a test gully transforms into the air bubbles taking various shapes, the received photometric levels behind the test gully with fresh water flow are affected by the light scattering through these agitating air bubbles. Thus, the normalized lumen level through the test gully at each test condition with fresh water is initially detected by present computerized optical system. The photometric receiver transmits the received light signal to the Personal Computer (PC), giving rise the lumen reference to determine the completion of self-depuration process at each test condition. By way of feeding the mixed water at the particular test condition defined by Re_L and α, namely the interfacial Reynolds number and void fraction of entry flow, the reference lumen levels at the pure water flow conditions (L_1) are pre-determined. It is interesting to note that, as the resolving air bubbles in the test gully reflect and scatter light, the instant lumen values at each pure-water test conditions oscillate about the corresponding L_1 reference. With self-depuration tests, the water trap stored in the test gully is dyed by the black ink to give the pre-defined lumen level (L_0) for a particular set of tests. The L_0 level at each "dark" test condition is controllable by adjusting the ink concentration and appears as a stable value due to the absence of air bubble prior to feeding the mixed water into the test gully. After charging the mixed water into the test gully, the instant lumen level (L) starts rising from L_0 toward L_1. The detailed temporal lumen (L) variation from L_0 to L_1 reflects the self-depuration performance for the test gully. For the test gullies with different geometries or different entry flow conditions, the L_0 and L_1 references are accordingly varied and measured. The temporal variation of normalized lumen in terms of L/L_1 is used to quantitatively characterize the self-depuration performance for each test gully. The time lapse taken for L/L_1 approaching 0.99 is defined as the self-depuration period correspond to the particular test condition.

Numerical Method and Simulation Details

With the Flow-3D code, the continuity equation and Navier-Stokes equation which describes the momentum conservation law for incompressible viscous flow within the fluid domain Ω surrounded by a piecewise smooth boundary Γ are described by Equations (1) and (2) respectively:

$$\nabla \cdot u = 0 \tag{1}$$

$$\frac{\partial u}{\partial t} + (u \cdot \nabla)u = -\nabla p + \frac{1}{Re}\nabla^2 u + f \tag{2}$$

In Equations (1) and (2), u, p, f, Re, t respectively denote the fluid velocity vector, pressure, additional force source terms, Reynolds number and time.

The solution in Ω domain satisfies the initial condition of $u = u_0$ and the non-slip boundary conditions on the solid boundary Γ. The geometries for numerical simulations are identical with the experimental test models using the scaling factor of unity as shown by Figure 1c. The origin of present *XYZ* coordinate system locates at the center of the bottom plate. Within the calculation Ω domain, the numerical solutions are obtained using the fine grid cells of length 1.5 mm. The air pressures for the voids in the water stream are assumed as 1.013×10^5 Pa (1 atm). Flow entry conditions for both gullies with and without SGV are identical with the total discharging rate of 30 L/min. For each entry port, the water flow rates, Q_A and Q_B, are set at $Q_A = Q_B = 15$ L/min with the void fraction of unity. This numerical study simulated the temporal variations of the interfacial air-water flow structures, including the 3-D distributions of *Fr*, vorticity and static pressure, for disclosing the complex two-phase flow phenomena in the test gullies without and with SGV. For the present numerical model, the intensity of non-linearity and convective effects are sensitive to the magnitude of volume flow rate from each inlet. The Sommerfeld radiation boundary conditions are selected as the outlet flow boundary conditions so that the study for the effects of wave interactions with the solid surfaces is permissible. Justified by the experimental observations, the lumped bubbly flows are selected as the interfacial flow structures throughout the calculations.

RESULTS AND DISCUSSION

Flow Structures

For establishing the comparative reference results, the flow structures in the test gully without SGV are detected against which the flow structures detected from the test gully with SGV are compared to disclose the SGV impacts on the hydrodynamic performances. The basic flow structures identified from the flow snapshots detected at all the tested flow rates (Q) of 10, 20, 30 and 40 L/min with single and twin entry flows remain similar for each type of test gullies. The basic flow structures in the test gullies without and with SGV are comparatively presented in Figure 2 at the maximum discharging rates. Having charged the mixed water from the twin entry ports, an entry vortex is formulated to convect the downward air-water stream into the primary drum. Justified by the convex curvature along the free surface of the entry vortex, the regional hydrodynamic performances for this type of test gullies are governed by the free vortex flow. However, near the center of the entry vortex, the contour of vortex reverts to be concave, featuring the forced vortex. The entry vortex in the primary drum is thus a mixed vortex. After the downward vortical air-water stream impinging onto the base plate of the test gully, the radially spreading air-water flow turns to be up-lifted through the 180°. sharp bend into the annular pathway between primary and secondary drums. Air bubbles entrained by the entry vortex are formed and drifting in this annular flow pathway, emerging the noticeably differential air-water interfacial activities between the tested gullies without and with SGV as compared by Figure 2. Clearly, the near-wall flows tripped by the angled SGV stratify the air bubbles to drift in the direction along the SGV orientation. In the test gully without SGV, the chaotic interactions among the up-drifting air bubbles take place in the annular passage, triggering considerable flow instabilities to amplify the air-pressure oscillations above the free surface between the primary and secondary drums. With the stabilized air-bubbles drift in the annular pathway among the upward flows for the test gully with SGV, the maximum discharging rates at present pressure potentials tested are increased more than 7% from those through the test gully without SGV.

Numerical simulations successfully capture all the dominant flow structures detected by the experimental study for the test gullies with and without SGV. The numerical test results obtained at water inflow rate for each entry port at 15 L/min show favorable agreements with the experimental measurements, thus confirming the calculated flow and pressure fields at the air-water flow conditions. The distributions of instant fluid velocity and pressure over the middle vertical planes of $Y = 0$ and $X = 0$ at $t = 10$, 20 and 30 s with $Q_A = Q_B = 15$ L/min are collected in Figure 3. In primary drum and the annular pathway between primary and secondary drums, the typical gravity-driven hydrostatic pressure variations are observed. When the upward air-water stream spills out of the annular pathway, the radial spreading water screen emitted from the top rim of the secondary drum envelops air bubbles. The free surface surrounding the outer wall of the secondary drum takes the unsteady wavy pattern for both gullies as shown by Figure 3. In the annular pathway between the primary and secondary drums and at the wavy free surface outside the secondary drum, the agitating bubbly air-water flows formulate the unstable flow region in this type of gully. Except in the agitating bubbly air-water flow region among which the air-bubble drifts are considerably affected by SGV as seen in Figure 2, the air-water flows in the gullies with and without SGV as shown by Figure 3 share the similar pattern. Many small-scale vortices with short life cycles are intermittently developed and resolved in both gullies with and without SGV.

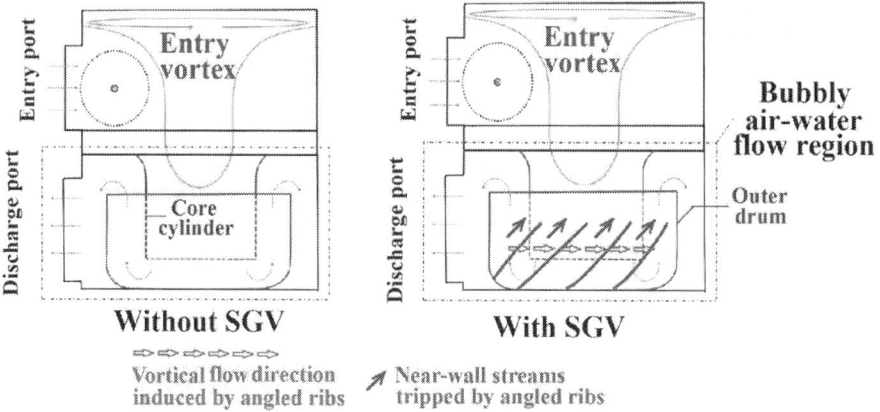

Figure 2. Air-water flow structures in test gully without SGV at $Q_A + Q_B = 65$ L/min and in test gully with SGV at $Q_A + Q_B = 70$ L/min.

To depict the complex unsteady air-water flow structures in present gullies without and with SGV, the three dimensional distributions of instant Froude number (Fr) at $t = 5$ and 30 s are calculated and collected in Figure 4. Present Fr is defined as the ratio of fluid velocity to the gravitational wave velocity to physically respond the ratio of inertial to gravitational forces for indicating the relative resistances of submerged air bubbles moving through the water stream. As compared with Figure 4, the Fr levels among the agitating bubble flow region in the gully without SGV are higher than the counterparts in the gully with SGV. Even with the protruding SGV to add the associated frictional and form drags along the flow pathway in the gully with SGV, the flow resistances attributed to the chaotic air bubble agitations in the gully without SGV still supersede the additional flow resistances added by the SGV; which leads to the increased maximum flow rates under the same pressure heads from the discharges for the gully with SGV. In Figure 4, the complete 3-D flow structures formulated by the entry vortex, agitating bubbly flow region along the serpentine flow pathway and the discharge flow with unsteady wavy free-surface are similar for both gullies without and with SGV to signify the characteristic flow pattern for this type of gully.

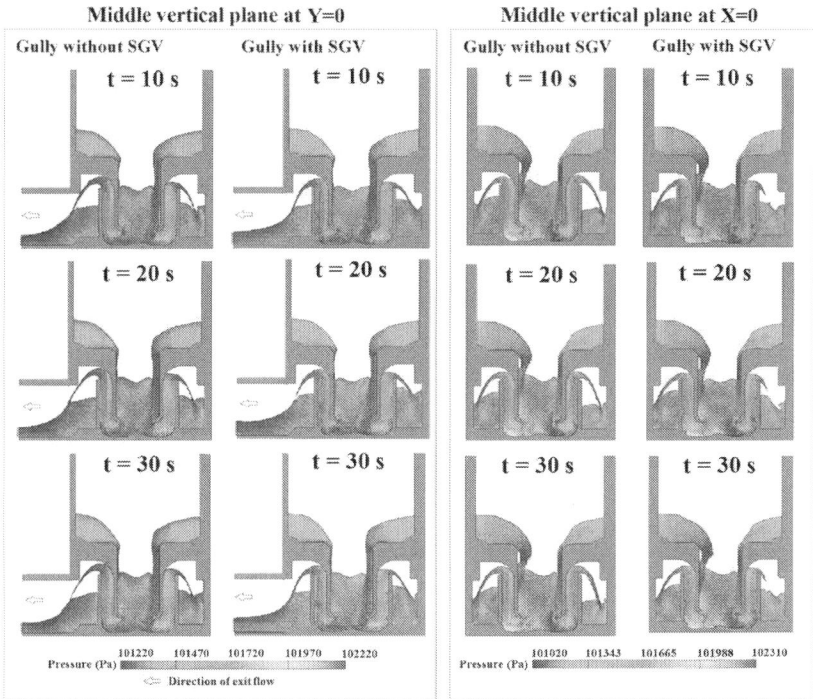

Figure 3. Distributions of instant fluid velocity and pressure over middle vertical planes of Y = 0 and X = 0 at t = 10, 20 and 30 s with $Q_1 = Q_2 = 15$ L/min.

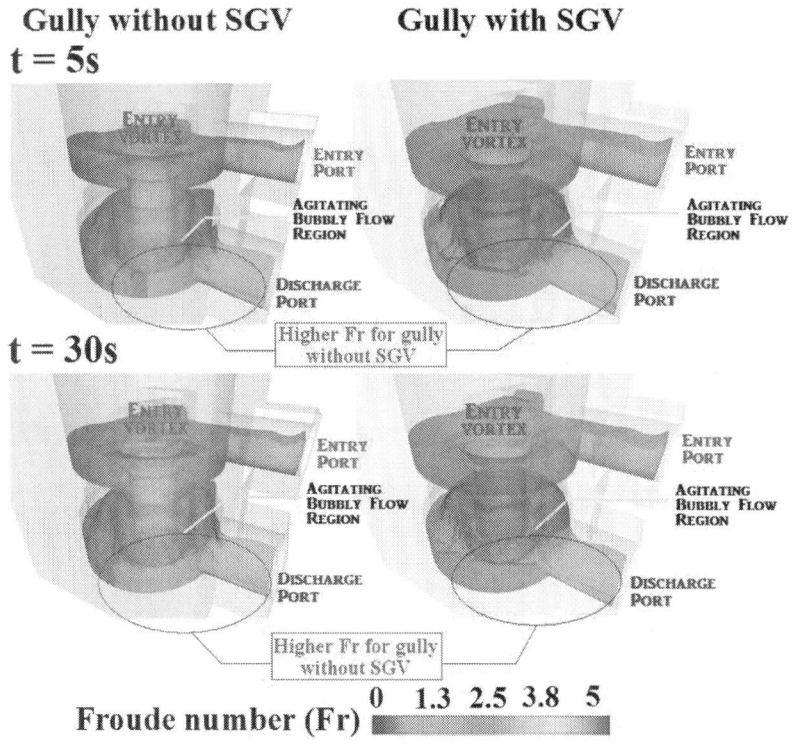

Figure 4. Three dimensional distributions of instant Froude number reflecting the overall flow structures in gullies with and without SGV.

Air Entrainments by Entry Vortex

For this type of gully, the downstream air-water flow structures are affected by the flow phenomena caused by the entry vortex, which include the considerable air entrainments. Following the conventional vortex theory, considerable radial pressure variations over the free surface and among the vortex are generated and affected by local fluid velocities. This is demonstrated by Figure 5, which compares the distributions of instant velocity and pressure contours between the gullies with and without SGV over three horizontal XY planes at $Z = 22$, 34 mm that are sectioned through the annular pathway between the primary and secondary drums and at $Z = 46$ mm under the primary drum. As Z increases, the gravitational effect increases the hydrostatic pressures in general, which is evidently shown by sequentially examining the three pressure contours obtained at $Z = 22$, 34, 46 mm at each t selected shown by Figure 5. At $Z = 22$ mm, the XY section through the exit port is fully occupied by the airflow; whereas the evident anti-clockwise vortex circulation are already emerged to fully occupy the primary drum. At $Z = 34$ mm, the pressures along the vortex outer edge are further elevated but moderated at $Z = 46$ mm. When the downward vortex stream is radially spread on the XY plane at $Z = 46$ mm, the characteristic signatures for vortex are according weakened for both gullies as demonstrated by Figure 5. With all the flow fields sectioned through the XY planes at $Z = 22$, 34 and 46 mm, the vortex core consistently show the lowest pressure levels due to the high fluid velocities. As the fluids approach the center of vortex, the increased fluid velocities are accompanied with the reduced static pressures. Once the static pressures over the free surface of the entry vortex fall less than the atmospheric level, the surrounding air above the entry vortex is entrained into the swirling liquid pool and converted to the air-bubbles by the surface tension effect. With the air entrainments by the entry vortex, a considerable amount of drifting air bubbles in the flow pathways is consistently observed even if the void fraction (α) over the flow entry ports is zero at the a full-water conditions. Although the resolving air bubbles in the present test gully are partially attributed to the local pressure reductions along the flow pathway, the air entrainment by the entry vortex is considered as the manifesting mechanism responsible for introducing air bubbles into the water stream. This is demonstrated by Figure 6 in which a series of continuous flow snapshots are selected to illustrate the process of air entrainment by the entry vortex.

To experimentally verify and visualize the mechanisms for the air entrainment by the entry vortex, the temporal variations of the airflow pressures, starting from charging the mixed water into the test gully, are individually detected at the various Z locations along the vertical central core ($X = Y = 0$) as depicted in Figure 6a. At $Z = 74$ mm, the probe of pressure sensor is about 1 mm above the liquid surface of the entry vortex-core. All the temporal variations of the airflow static pressures collected in Figure 6a from the different Z locations follow a similar varying trend. Within an initial period about 30 s after feeding mixed water into the test gully at the single entry condition of $Q = 30$ L/min, the entry vortex remains as developing; whereas the liquid level in the gully is up-rising to compress the trapped air within the gully drum, leading to the positive pressure heads along the central core as shown by Figure 6a. At the instant that the discharge of mixed water flow is partially choked, the upstream pressure waves generate an abrupt pressure increase at all the measured Z locations as shown by Figure 6a. Followed by the sudden airflow pressure rises shown by Figure 6a, the growing strength of the entry vortex keeps accelerating and dragging the airflow adjacent to the free surface of the entry vortex, leading to the subsequent reducing trend of pressure reductions at all the Z locations seen in Figure 6j. The negative airflow pressures at the locations close to the free surface of entry vortex are then emerged and stayed to trigger the process of air entrainment as demonstrated by the following Figure 6b–j. Due to the complex and interactive air-water interfacial mechanisms among the vortex core region, the static airflow pressures start oscillating about the atmospheric level to promote the unsteady air entrainments by the entry vortex as $t > 70$ s for this particular test condition.

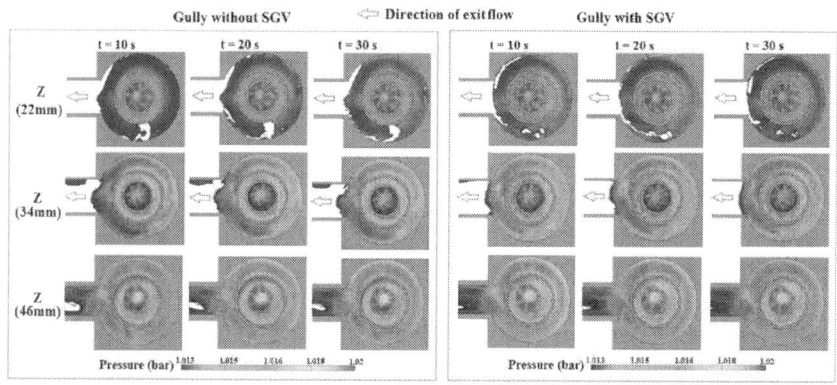

Figure 5. Distributions of instant velocity and pressure contour for gullies with/without SGV over horizontal XY planes at $Z = 22, 34, 46$ mm.

The process of vortex deformation is initially observed at instants seen in Figure 6b–c by sharpening the vortex core in downward direction seen in Figure 6c. As a result of the driven pressure gradients on the free surface of the entry vortex, a lumped air bubble is formulated at the vortex core; but still coherently attached on the free surface of the entry vortex as shown by Figure 6d. After a short time lapse, the separation of air bubble into the liquid pool is observed as seen in Figure 6e; which can be occasionally followed by another sequence of vortex-core deformation and air-bubble separation seen in Figure 6f. The large-scale separated air bubble that submerges into the swirling liquid pool is generally broken into small air bubbles which scatter underneath the vortex core as indicated by Figure 6g–h. The interfacial air-bubble evolutions disclosed by sequentially viewing the flow snapshots detected at the instants shown by Figure 6b–h are followed by the subsequent vortex-core deformation as typified in Figure 6i to complete an air-entrainment process induced by the entry vortex. The successive process for another air entrainment is initiated with the flow image shown by Figure 6j. It is noticed, with present test gullies, the entire air entrainment process by entry vortex, as typified by Figure 6b–j, is completed within 1 s.

In addition to the considerable flow resistances by the air bubbles in the flow passages formulated in the gullies without and with SGV as demonstrated by Figure 4, the entrained air into the water stream also affect the vorticity distributions in the gullies. To explore the impact of entrained air on vorticity distributions, the instant vorticity contours for the gullies with/without SGV over horizontal XY planes at Z = 22, 34, 46 mm at t = 10, 20 and 30 s with $Q_A = Q_B$ = 15 L/min, which corresponding to the Computational Fluid Dynamics (CFD) scenarios collected in Figure 3, are compared by Figure 7.

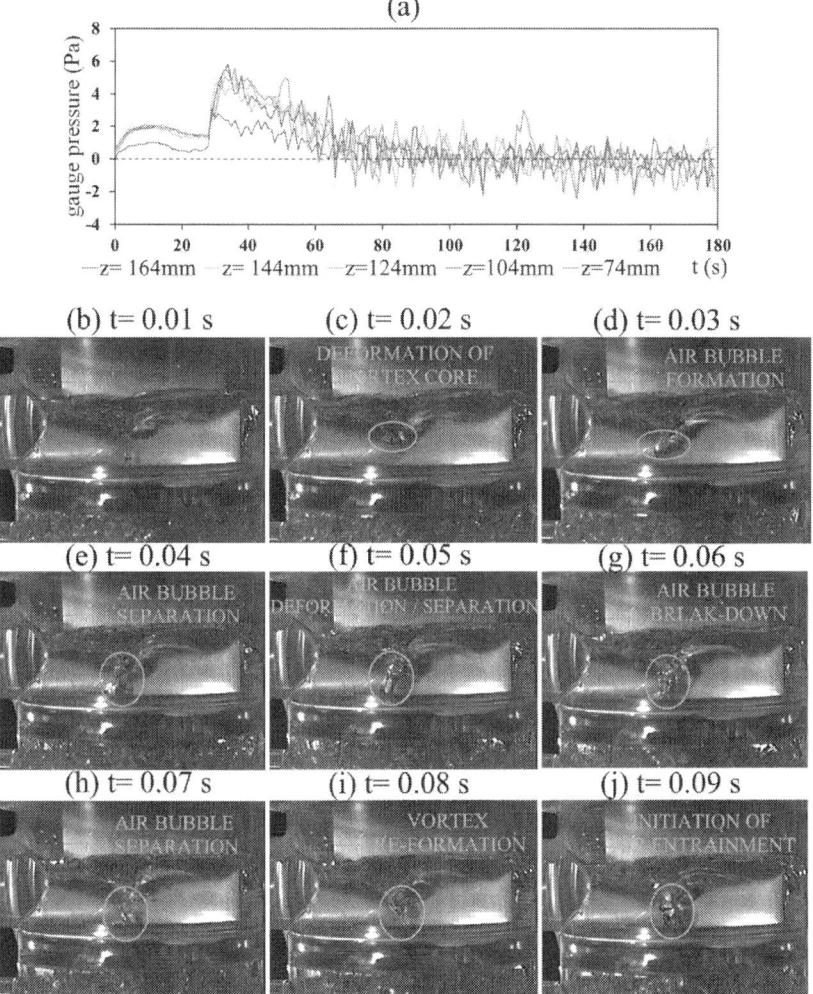

Figure 6. Temporal airflow pressure variations and corresponding flow snapshots demonstrating the process of air entrainment by entry vortex.

It is interesting to note the ring of high vorticity circling around the center of entry vortex. Due to the air-entrainment taking place at the center of the entry vortex, the development of local angular momentum by the shearing action resulting from the particular fluid velocity field is interfered. As a result, the local vorticity at the center region of the entry vortex is weakened to be less than those emerging along the surrounding rim shown by Figure 7. Over the annular sections between the gully casing and the secondary drum, several spots show the negative vorticites, in particular along the air-water interfacial boundaries marking as the black solid lines in Figure 7. The counteracting circulations for the air bubbles in the water stream are suggested by present numerical results. Above all, with applications to drainage systems, present type of gullies can be classified as the appliance capable of entraining air into the drainage system. Flow instabilities are mainly attributed to the air bubble interactions in the agitating bubbly flow region specified by Figure 4.

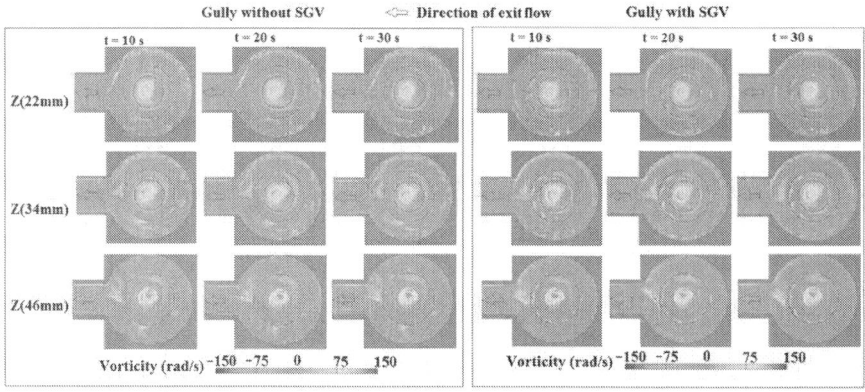

Figure 7. Distributions of instant vorticity for gullies with/without SGV over horizontal XY planes at $Z = 22, 34, 46$ mm.

Air Bubble Drifts in Test Gullies with/without SGV

As the primary contributions of present SGV for improving the hydrodynamic performances of this type of gullies, the near-wall water streams tripped by the angled SGV assist to guide the drifting air bubbles over the agitating bubbly flow region. This is demonstrated by Figure 8 in which the trajectories of drifting air bubbles in the test gullies without and with angled SGV are compared. The instant flow snapshots adopted to identify the drifting trajectories for the air bubbles in the agitating bubbly flow region are also shown in Figure 8. As summarized in the conceptual flow diagram for the test gully without SGV in Figure 8, the drifting trajectories of air bubbles mainly follow three routes indicated by the A, B,

C traces in the flow snapshots as shown in Figure 8. Along the drifting routes A and B in the test gully without SGV, the complex bubble collisions and coalescences and oscillations are observed. Relative to the gully with SGV, the highly agitated free surface between the secondary drum and the gully outer cylindrical casing is observed for the test gully without SGV. By fitting the angled SGV along the cylindrical wall of the secondary drum, the air bubbles are drifting along with the near-wall water streams tripped by the angled SGV so that all the A, B, C trajectories for air bubble drifts in the agitating bubble flow region are guided/stratified along the angled SGV direction to moderate the flow instabilities caused by the random air bubble collisions and coalescences and oscillations.

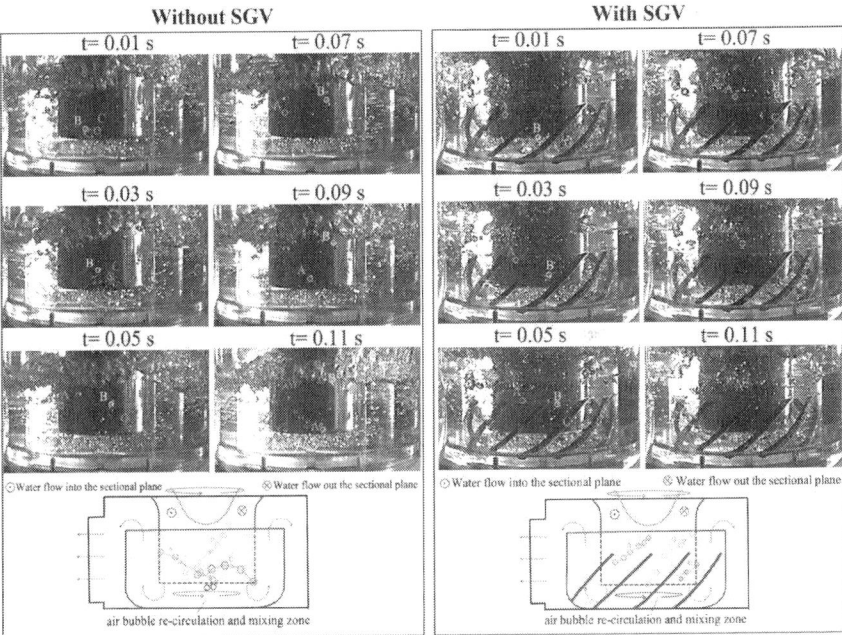

Figure 8. Drifting trajectories of air bubbles in test gullies without/with SGV.

Self-Depuration Performances

While the agitations of air-bubbles and the motion of free-surface in present test gullies without and with SGV are considerably different, the performances of self-depuration are similar. Figure 9 compares the temporal variations of L/L_1 ratios at all the tested flow rates with single and twin flow entry conditions for the test gullies without and with SGV. As compared by Figure 9, the temporal variations of L/L_1 ratios at all the tested flow conditions with single and twin flow entries follow the similar pattern. Prior to charging the mixed water into each test gully, the dye concentration is controlled to provide the referenced L/L_1 ratios at about 0.4. After feeding the mixed water into each test gully, an initial start-up period with stable L/L_1 levels proceeds about 10 s. Following the stable period with L/L_1 ratios at about the reference "dark" condition, the L/L_1 ratios increase sharply within a short period about 5 s. The physical implication for such rapid L/L_1 increase is the significant improvement for the self-depuration performance attributed to the development of entry vortex which effectively discharges the dyed water and replenishes with the supplied fresh mixed water. While the air entrainment is mainly caused by the entry vortex, the self-depuration performance is considerably improved by the entry vortex which rapidly replaces the dyed water by the mixed water. After the period of rapid L/L_1 increase, an exponential-like period of moderate L/L_1 increase over the period about 3–10 s is followed. As Q increases, the initiation of the rapid L/L_1 increase is advanced as shown by Figure 9. Thus, the consistent reduction of self-depuration period by increasing the discharge capacity is observed in Figure 9. The variations of self-depuration time for each test gully against total entry water flow rate at single and twin entry conditions are summarized in Figure 10. As compared with the three data trends obtained at single and twin flow entry conditions, the self-depuration time for the test gullies without and with SGV at the two single flow entry conditions labeled as QA and QB in Figure 10 are similar. By feeding the air-water mixed flows from present two perpendicular flow entry pipes in tangent with the gully drum, the flow momentums required to formulate the entry vortex are likely to be raised from the conditions with single entry flow. With the enhanced vortical strength for the entry vortex at the twin-entry flow conditions, the self-depuration time is consistently less than the single-entry counterpart for the test gullies without/with SGV as shown by Figure 10. Justified by the data trends revealed in Figure 10, the empirical correlations for the self-depuration time are devised as Equations (3)–(5) and (6)–(8) for present test gullies without and with gullies:

$T = -14.38 \ln(Q_A) + 73.77$ (single flow entry A for gully without SGV)

$T = -13.83 \ln(Q_B) + 73.33$ (single flow entry B for gully without SGV)

$T = -15.01 \ln(Q_A + Q_B) + 72.75$ (twin flow entry A+B for gully without SGV)

$T = -13.19 \ln(Q_A) + 70.35$ (single flow entry A for gully with SGV)

$T = -13.1 \ln(Q_B) + 68.81$ (single flow entry B for gully with SGV)

$T = -12.55 \ln(Q_A + Q_B) + 63.37$ (twin flow entry A+B for gully with SGV)

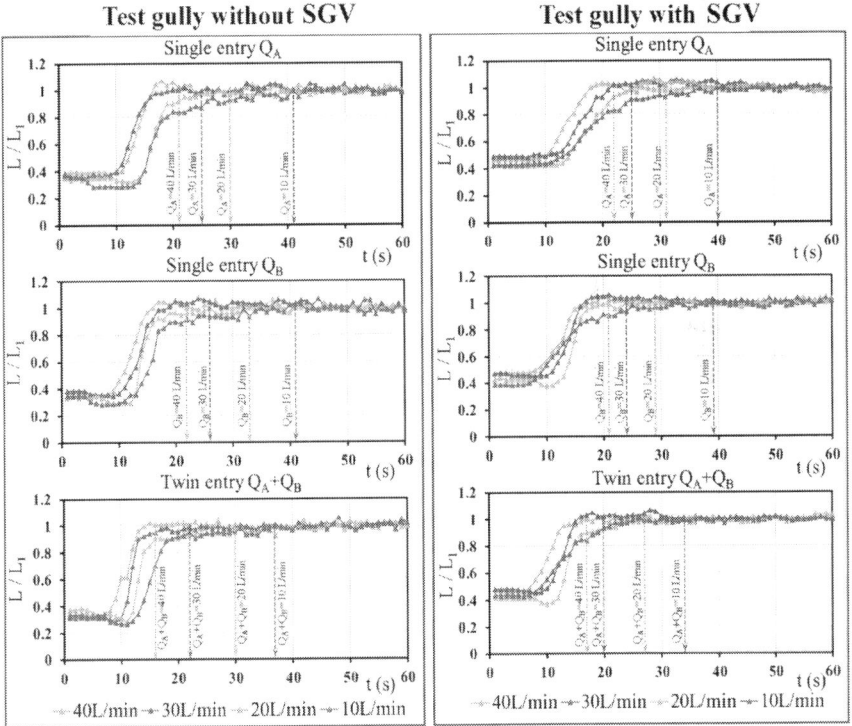

Figure 9. Temporal $L/L1$ variations for test gullies without/with SGV at single/twin entry conditions.

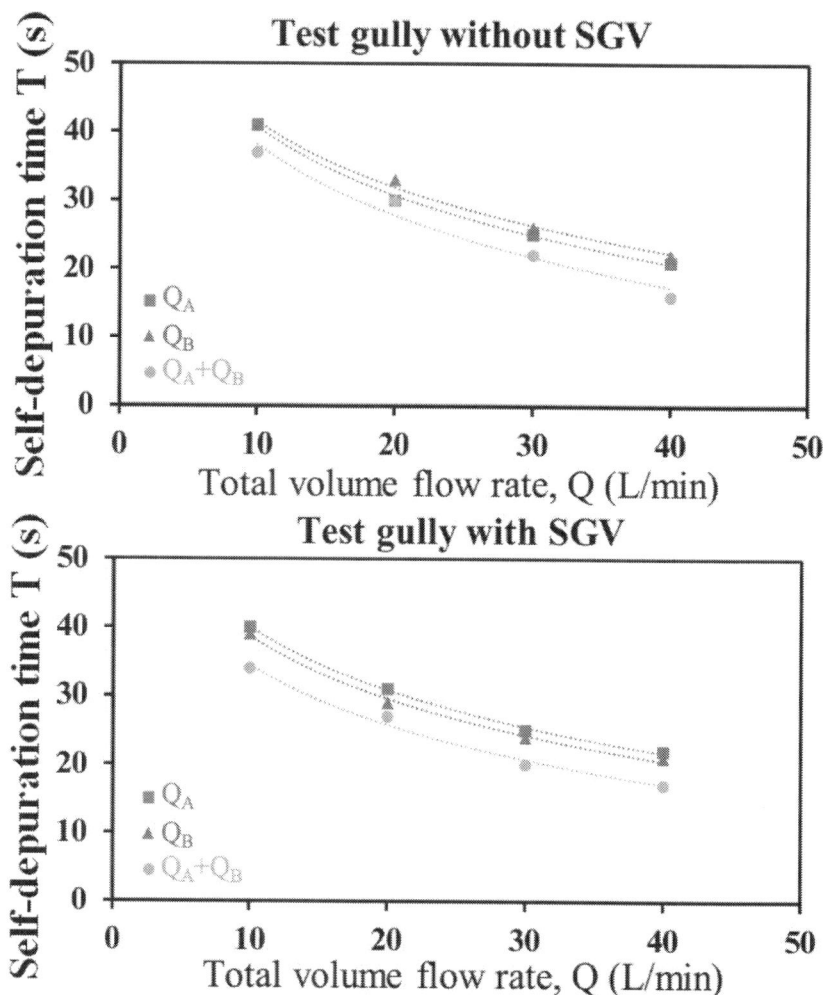

Figure 10. Variations of self-depuration time for test gullies without/with SGV at single/twin entry conditions.

CONCLUSIONS

This experimental and numerical work comparatively examined the hydrodynamic performances of two test gullies without and with SGV to enlighten the air-water flow structures, air entrainment mechanisms, air-bubble drifts and self-depuration properties. The conclusions emerge from this study are served as the design considerations with the applications to drainage systems in buildings. With the entry vortex formulated in the primary drum; this type of gullies is classified as the appliance that entrains air into the drainage system.

Air bubbles entrained by the entry vortex in present test gully without SGV interact chaotically in the agitating bubbly flow region. With SGV on the cylindrical wall of test gully, the near-wall flows tripped by the angled SGV stratify the drifting trajectories of the air bubbles, leading to the stabilized air-bubble interactions. Justified by the 3-D Fr distributions, the flow resistances attributed to the chaotic air bubble agitations in the gully without SGV supersede the flow resistances caused by the SGV. The maximum discharging rates for the test gully with SGV at present pressure head of 1.2 m water-height are increased more than 7% from the discharges by the gully without SGV.

After an initial short period of stable low L/L_1 levels during which the entry vortex is under development, the rapid L/L_1 increase followed by the exponential-like moderate L/L_1 increase reflects the characteristic self-depuration property for this type of gullies with entry vortex. The consistent reduction of self-depuration period by increasing the discharge capacity is consistently observed for present test gullies without and with SGV. Two sets of empirical correlations that permit the estimation for self-depuration periods at single and twin entry flow conditions for present test gullies without and with SGV are devised to assist the relevant industrial applications.

AUTHOR CONTRIBUTIONS

Der-Chang Lo performed the Computational Fluid Dynamics (CFD) simulation to reveal the fundamental air-water flow phenomena in the multi-entry gully. Jin-Shuen Liou carried out the experimental tests and data analysis. Shyy Woei Chang wrote the paper, proposed the SGV and formulated the experimental method to investigate the effects of SGV on the hydrodynamic performances for the multi-entry gully.

REFERENCES

1. Swaffield, J.A.; Campbell, D.P. Air pressure transient propagation in building drainage vent systems, an application of unsteady flow analysis. *Build. Environ.* **1992**, *27*, 357–365.

2. Swaffield, J.A.; Campbell, D.P. The simulation of air pressure propagation in building drainage and vent system. *Build. Environ.* **1995**, *30*, 115–127.
3. Chang, S.W.; Hsieh, C.-M.; Lin, C.Y.; Liou, H.-F. Air-water drainage flow through finned bend. *J. Asian Archit. Build. Eng.* **2012**, *11*, 177–184.
4. Swaffield, J.A.; Jack, L.B.; Campbell, D.P. Control and suppression of air pressure transients in building drainage and vent systems. *Build. Environ.* **2004**, *39*, 783–794.
5. Cheng, C.L.; Mui, K.W.; Wong, L.T.; Yen, C.J.; He, K.C. Characteristics of air pressure fluctuations in high-rise drainage stacks. *Build. Environ.* **2010**, *45*, 684–690.
6. Swaffield, J.A.; Campbell, D.P. Numerical modeling of air-pressure transient propagation in building drainages, including the influence of mechanical boundary conditions. *Build. Environ.* **1992**, *27*, 455–467.
7. Ma, J.; Oberai, A.; Lahey, R., Jr.; Drew, D. Modeling air entrainment and transport in a hydraulic jump using two-fluid RANS and DES turbulence models. *Heat Mass Transf.* **2011**, *47*, 911–919.
8. Ma, J.; Oberai, A.; Drew, D.; Lahey, R., Jr.; Hyman, M. A comprehensive sub-grid air entrainment model for RaNS modeling of bubbly flows near the free surface. *J. Comput. Multiph. Flows* **2011**, *3*, 41–56.
9. Wright, G.B.; Swaffield, J.A.; Arthur, S. The performance characteristics of multi-outlet siphonic rainwater systems. *Build. Serv. Eng. Res. Technol.* **2002**, *23*, 127–141.
10. Wright, G.B.; Arthur, S.; Swaffield, J.A. Numerical simulation of the dynamic operation of multi-outlet siphonic roof drainage systems. *Build. Environ.* **2006**, *41*, 1279–1290.
11. Chang, S.W.; Lo, D.C. Chapter 6 Air-Water Two-Phase Flows with Applications to Drainage System. In *Advances in Multiphase Flow and Heat Transfer*, 2nd ed.; Bentham Science Publishers Ltd.: Sharjah, UAE, 2009; pp. 176–215.
12. Chang, S.W.; Lo, D.-C.; Liou, H.-F.; Liou, J.S. Hydrodynamic performances of gully with air-water flows in drainage system. *J. Water Resour. Supply Drain. Build.* **2014**, *1*, 1–19.

CITATION

Der-Chang Lo, Jin-Shuen Liou and Shyy Woei Chang, Hydrodynamic Performances of Air-Water Flows in Gullies with and without Swirl Generation Vanes for Drainage Systems of Buildings, doi:10.3390/w7020679.

CHAPTER 6

Flexible Heat Pipes with Integrated Bioinspired Design

Chao Yang, Chengyi Song, Wen Shang, Peng Tao, , Tao Deng,

State Key Laboratory of Metal Matrix Composites, School of Materials Science and Engineering, Shanghai Jiao Tong University, Shanghai 200240, China

ABSTRACT

In this work we report the facile fabrication and performance evaluation of flexible heat pipes that have integrated bioinspired wick structures and flexible polyurethane polymer connector design between the copper condenser and evaporator. Inside the heat pipe, a bioinspired superhydrophilic strong-base-oxidized copper mesh with multi-scale micro/nano-structures was used as the wicking material and deionized water was selected as working fluid. Thermal resistances of the fabricated flexible heat pipes charged with different filling ratios were measured under thermal power inputs ranging from 2 W to 12 W while the device was bent at different angles. The fabricated heat pipes with a 30% filling ratio demonstrated a low thermal resistance less than 0.01 K/W. Compared with the vertically oriented straight heat pipes, bending from 30° up to 120° has negligible influence on the heat-transfer performance. Furthermore, repeated heating tests indicated that the fabricated flexible heat pipes have consistent and reliable heat-transfer performance, thus would have important applications for advanced thermal management in three dimensional and flexible electronic devices.

INTRODUCTION

Human as a natural three dimensional flexible system has integrated many natural flexible accessories. One of such flexible accessories is the blood vessel, which serves not only as a transportation highway for nutrition delivery, but also as a flexible thermal conductor for thermal management of human bodies. This paper intends to build an artificial flexible heat conductor using the principle of heat pipes. As a passive heat-transfer device with high effective thermal conductivities and requiring no maintenance, heat pipes have become an important tool for thermal management of electronic systems[1]. A heat pipe normally consists of a hermetic container, working fluid that can absorb heat from the heat source upon vaporization and release heat at the condenser section, and a wick structure that supplies capillary forces to pump the condensed liquid back to the hot evaporator section forming a heat-transfer cycle loop. The container materials were mostly made of rigid metals such as copper, aluminum and stainless steel, due to their high thermal conductivity, mechanical robustness and excellent barrier properties. Thus, the fabricated heat pipes are usually used in a straight configuration, lacking the flexibility and bendability to fulfill their wider applications required in contorted configurations where heat sources and heat sinks are not in the same plane or the heat source is not stationary [2] and [3]. A flexible heat pipe similar to the flexible human blood vessel is in urgent need in order to meet the thermal management requirement of three dimensional and flexible electronic systems.

The earliest effort in flexible heat pipes can be dated back to 1970. Bliss et al. [4] connected the rigid copper evaporator and condenser with a brass bellowed tube in the adiabatic section and realized bending of the heat pipes at 45° and 90° during horizontal operation. While bellows improved the bendability, in most cases the heat pipes were still rigid and were integrated into systems with bends fabricated a priori [5] and [6]. Heat pipes that can be easily bent for many times and easily bent into designed geometries are highly desired. In recent years, polymer based heat pipes have attracted increasing attention owing to the advantageous features from polymers including flexibility, lightweight, good processability and low cost. Various polymers such as polypropylene [7], polyimide [8], liquid-crystal polymers [9], polyethylene terephthalate [10],[11] and [12] and silicone [13] and [14] have been utilized as the container material. These polymer based heat pipes were often flat, micro or pulsating heat pipes with small power capacities and large thermal resistances due to the low thermal conductivity of the polymer casing materials. To mitigate the issue of large thermal resistance for the pure polymer heat pipes, high thermal conductivity copper-filled thermal vias [9], copper mesh[11], or copper sheet [13] were fabricated in the evaporator and condenser sections. However, the fabrication usually involves complicated and time-consuming micro-fabrication processes such as multi-step molding, bonding and assembly.

Wick structure is another important component for designing and fabricating flexible heat pipes. Conventional sintered copper powders [1] or recently reported sintered copper felts [6] and [15] within copper tubes are not compatible with the low-temperature processing of polymer heat pipes. By contrast, copper meshes or copper fibers which were sintered at high temperatures under protective inert gas environment are used as the popular wicking materials for polymeric flexible heat pipes [16]. Besides the good bendability of the copper mesh, the wicking materials should have good wettability with working fluids to maximize their capillary pumping force. Previously, Oshman et al. [11]deposited an Al_2O_3/SiO_2 bi-layer coating on to the sintered copper mesh by atomic layer deposition technique to promote wettability of the mesh with water. Biological materials such as moss and *Rhacocarpus purpurescens* leaves have developed natural hierarchical micro/nano-structures rendering their surfaces superhydrophilic [17]. It is expected that by fabricating nature inspired hydrophilic materials and using them as the wicking material would be able to effectively improve the capillary pumping capabilities of heat pipes.

In this work, we report a facile approach to prepare flexible heat pipes that have integrated the bioinspired superhydrophilic wick structure and the flexible connector design in the adiabatic section. The cylindrical heat pipes were fabricated by connecting copper tubular evaporator and condenser with flexible polyurethane and using water as the working fluid. Superhydrophilic flexible strong-base-treated copper meshes were adopted as the wicking materials. Thermal performance tests were carried out with different thermal power inputs to the evaporator while the device was bent at 0°, 30°, 60°, 90° and 120°. Compared with the pure polymer heat pipes, our heat pipes demonstrated much smaller thermal resistances while maintaining excellent flexibility. It was found that bending has negligible impact on the thermal performance of the fabricated flexible heat pipes.

EXPERIMENTAL

Materials

Copper tubes, copper mesh (No. 300) and polyurethane tubes were purchased from Shanghai hydraulic pipe fittings Co., Ltd., Shanghai Hengxin wire & mesh Co., Ltd. and Shanghai Yihui Rubber & Plastics Co., Ltd., respectively. The bonding adhesive (TS1415) was purchased from Beijing Tianshan Kesaixin adhesive Co., Ltd. The HCl solution was obtained from Shanghai Lingfeng Chemical reagent Co., Ltd. KOH and $K_2S_2O_8$ were purchased from Sinopharm Chemical reagent Co., Ltd. and Aladdin reagent Co., Ltd., respectively.

Preparation of wick

The purchased copper meshes were subject to a chemical treatment by following the procedure described by Xie et al. [18]. The copper mesh was first immersed in a 4 mol/L HCl solution for 15 min followed by rinsing with deionized water. The acid-cleaned mesh was then transferred to the mixed solution of 0.065 mol/L $K_2S_2O_8$ and 2.5 mol/L KOH at 60 °C for 60 min. Finally, the treated mesh was cleaned and dried.

Fabrication of heat pipes

The cylindrical heat pipes were fabricated by utilizing copper tubes with an outer/inner diameter of 5/4 mm and a length of 150 mm at the evaporator and condenser section, and using a polyurethane tube with an outer/inner diameter of 8/5 mm and a length of 100 mm in the adiabatic section, respectively. The copper tubes were first cleaned by immersing them into 10 vol% sulfuric acid with the assistance of ultrasonic vibration. The copper tubes and the polyurethane tube were bonded together using an adhesive (TS1415) at room temperature for 24 h. The bonding was further mechanically strengthened by a tightening belt. Then, the aforementioned treated copper mesh was tightly inserted into the heat pipe to serve as the wicking material. Deionized water was selected as the working fluid. Three different filling ratios (20%, 30% and 40%) of deionized water were charged to identify the optimum filling ratio. The heat pipe was outgassed by heating the bottom section with a heater and a thermal couple was attached at the upper exit end to monitor the outgassing process. The charged heat pipe was clamped with a pinch-off tool and finally sealed with tungsten arc welding under the purge of Argon gas.

Characterization and property measurement

Microstructure of copper mesh wicking structures was examined by a scanning electron microscopy (FEI Sirion 2000). The wettability of the untreated and treated copper mesh wick was evaluated by measuring their contact angle with water.

Fig. 1 presents the test setup for heat-transfer performance measurement. The evaporator section was encased by a silicone rubber heater. The heater was connected to an adjustable DC power supplier (TPR 3005T, Shenzhen Atten Technology, Co. Ltd.). The condenser section was in close contact with a cooling plate (Shanghai Bilon Equipment) running with circulated chilling water at 15 °C. A porous plastic thermal insulation material was used to wrap the evaporator and adiabatic sections to prevent heat loss. Two K-type thermocouples (DM6801A, Nanjing Victor Equipment, Co.

Ltd.) were mounted on the heat pipe to record the temperature. The temperature readings were taken when they reached a steady state under a certain power input. The temperature difference between the evaporator (T_e) and the condenser (T_c) was used to calculate the thermal resistance (R) of the heat pipe as shown by the following equation:

$$R = \frac{T_e - T_c}{P}$$

where P is the heating power (from 2 W to 12 W). The heat pipes were tested in a straight configuration and bent with different angles (30°, 60°, 90° and 120°). An empty copper tube was used as the benchmark sample.

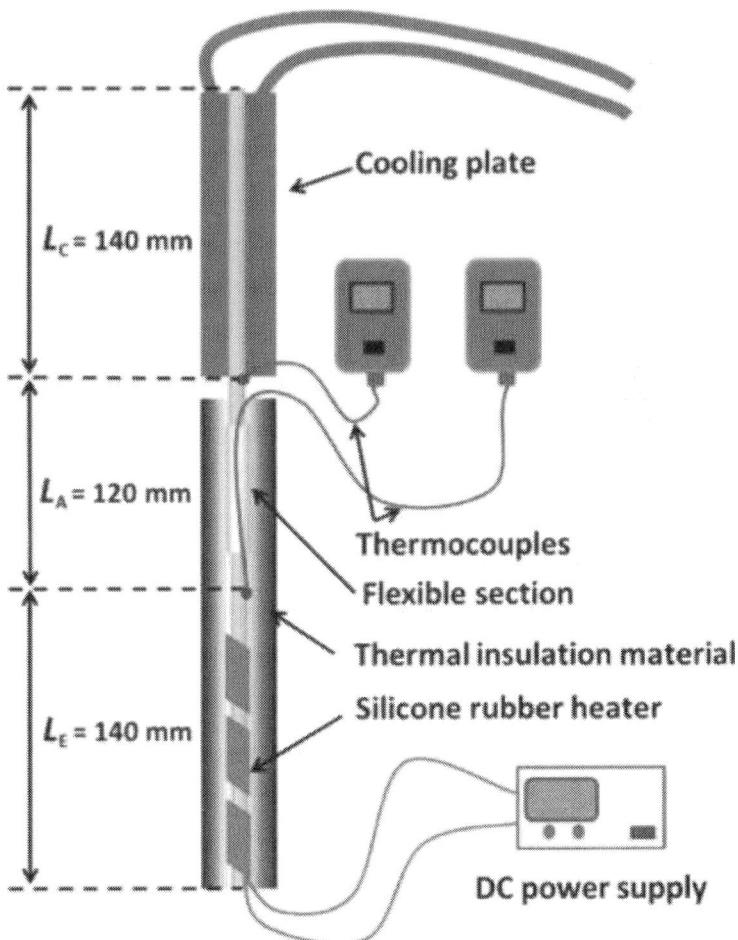

Figure 1. A schematic of test setup for measurement of the thermal resistance of flexible heat pipes.

RESULTS AND DISCUSSION

Fig. 2 presents the schematic and a photograph of the prepared flexible heat pipe. Considering that the total thermal resistance (R_t) can be roughly estimated as a sum of the thermal resistance of evaporator wall (R_e), vapor channel (R_v) and condenser wall (R_c), whole pure polymeric heat pipes would have a large R_t as polymers generally have a low thermal conductivity. Here we used polyurethane which is flexible but with a high thermal resistance only at the adiabatic section to decrease R_e and R_c. Compared with other polymer materials such as silicones, polyurethane has a good combination of flexibility, mechanical robustness and gas barrier properties. The metallic copper evaporator and condenser ensure the good heat-transfer capability of the prepared heat pipes.

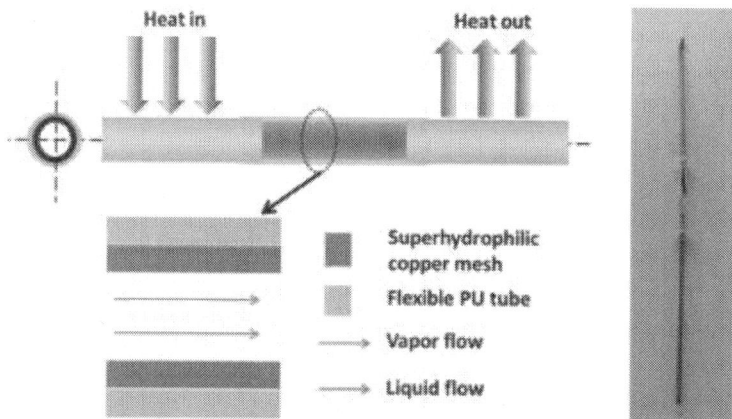

Figure 2. A schematic cross-section of the flexible heat pipe and a photograph of the fabricated heat pipe.

Fig. 3a shows a photograph of a curled copper mesh with a dimension of 30 mm×400 mm. The SEM image at low magnifications in Fig. 3b presents that the copper mesh has a wire diameter of ~36 μm and a spacing of ~45 μm. The porous copper meshes have been used as the wicking materials for flexible heat pipes as they are industrial available and robust enough to allow for repeated bending while providing a strong capillary pumping force. However, the purchased copper mesh is highly hydrophobic showing a contact angle of 135° even after cleaning with dilute sulfuric acid. To improve the wettability of copper meshes with the deionized water working fluid, Li and Peterson [16] sintered them at a high temperature of 1030 °C for 150 min under the purge of Argon gas. In addition, multi-layered copper meshes or copper mesh micro-groove hybrid wick structures [9] were designed and fabricated to improve wettability of the wick materials and thus to enhance their capillary pumping capability.

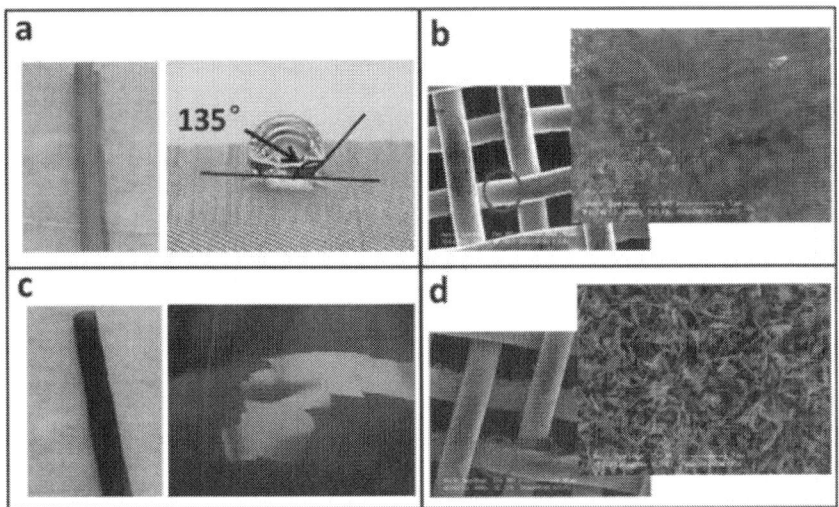

Figure 3. (a) A photograph showing a curled flexible hydrophobic copper mesh; (b) SEM images of copper mesh at low and high magnifications; (c) a photograph showing a flexible hydrophilic copper mesh after treatment with strong bases; and (d) SEM images of treated copper mesh at low and high magnifications.

As demonstrated by the photograph of a black curled mesh in Fig. 3c, after oxidation within a strong base solution the copper mesh turned into black colored but retained its flexibility feature. The treated mesh showed excellent wettability and water can easily spread on it. Comparison of the SEM images in Fig. 3c and d shows that unlike the smooth surface of the as-purchased copper meshes, dense acicular micro/nano-structures were observed on the surface of the treated copper mesh. According to description from Xie et al. [18], the involved chemical process is as following:

$$Cu+2KOH+K_2S_2O_8 \rightarrow Cu(OH)_2+2K_2SO_4.$$

After treatment, Cu was converted into hydrophilic $Cu(OH)_2$ and the hierarchical mico/nano-structured rough surface further amplified the wettability rendering a superhydrophilic wick structure. It should be mentioned that conventional sintering approaches often only generate hydrophilic wicks and require high-temperature processing. By contrast, the strong base oxidation method is much simpler and yielding superhydrophilic structures.

To identify the influence of working fluid filling ratio on the heat-transfer performance of the heat pipes, three different volumetric loadings (20%, 30% and 40%) of deionized water were charged. Two thermocouples were mounted onto the evaporator and condenser to monitor the temperature change of the heat pipes in a vertical gravity-assisted configuration. Fig. 4a presents that the evaporator temperature (T_e) and the condenser temperature (T_c) increased linearly with increasing power input to the evaporator section as the heat transfer dominantly relies on heat conduction mechanism. Accordingly, the thermal resistance (R) of the copper tube showed a nearly constant value (~8 K/W). Heat pipes with different filling ratios demonstrated different heat-transfer performances. At low power inputs, in general, there is no enough driving force to observe the heat pipe effect as evidenced by the large temperature difference between T_e and T_c at 2 W and 4 W. When the working fluid filling ratio was 20%, until the input power reached 10 W the temperature difference significantly reduced (Fig. 4b). With higher filling ratios, the near constant-temperature heat-transfer effect was observed when the heating power reached 6 W (Fig. 4c and d).

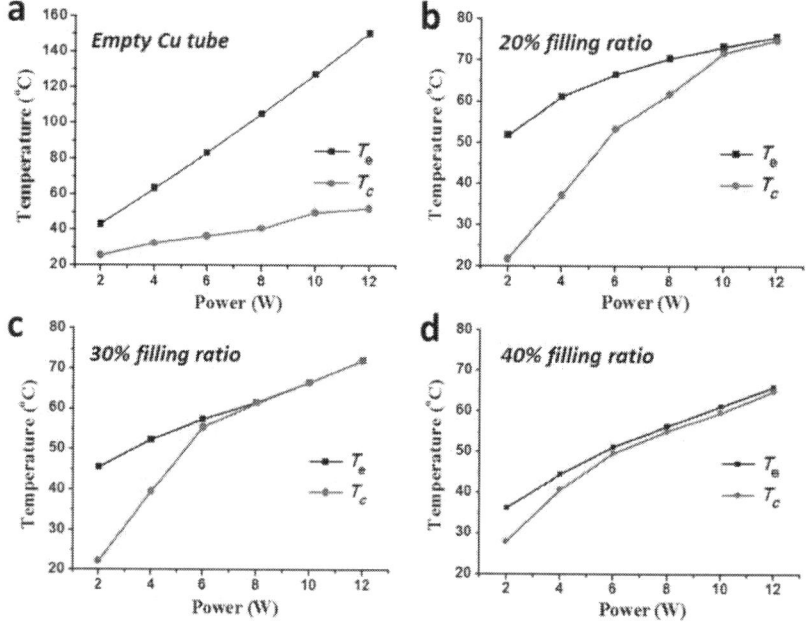

Figure 4. Temperature of the fabricated flexible heat pipe at evaporator and condenser sections: (a) empty copper tube; (b) 20% filling ratio; (c) 30% filling ratio; and (d) 40% filing ratio.

Fig. 5 presents that the thermal resistance of the heat pipes decreased with increasing power inputs. When the thermal input at the evaporator section was 2 W, the heat pipe filled with 40 vol% water even showed a lower thermal resistance than the benchmark copper tube sample. This probably could be related with the fact that with 40 vol% loading the upper liquid surface was very close to the thermocouple at condenser and water contributed as the additional heat conduction medium. By comparison, with 20% and 30% filling ratios the charged working fluid was not enough to fill the polyurethane tube. The low thermal conductivity polyurethane tube sacrifices the good thermal conducting properties of the fabricated heat pipe leading to a higher thermal resistance than the empty copper tube. Although 30% and 40% filling ratios yield similar thermal resistances, close examination indicates that the thermal resistance of the heat pipe charged with 30% filling ratio was even lower (less than 0.01 K/W). This value is much smaller than the data reported by others in pure polymeric heat pipes. It appears that 30% is an optimum filling ratio and this value is also close to the reported data in the literature [9], [10] and [11]. At low filling ratios, there was no enough working fluid delivering heat from evaporator to condenser. The evaporator temperature kept increasing rapidly leading to a large temperature difference between T_e and T_c. While at high filling ratios, the heat at the evaporator initially can be delivered to the condenser, but too much working fluid would over-flood the evaporator section resulting in decreased wicking forces to pump the condensed fluid back to evaporator [19]. This finally leads to rising temperature at the evaporator and thus large temperature differences and thermal resistances.

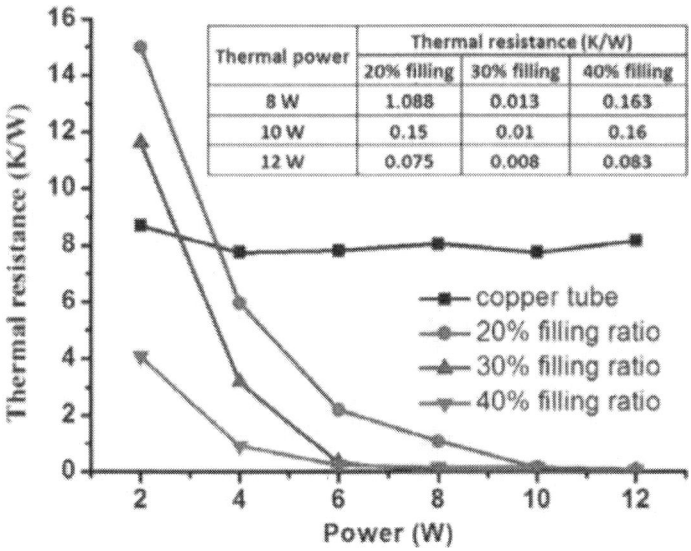

Figure 5. Thermal resistance of fabricated flexible heat pipes charged with different filling ratios of working fluid.

In order to evaluate heat-transfer performances in the bending condition, the thermal resistance of heat pipes filled with 30% water was tested at different bending angles from 30° to 120° as schemed by Fig. 6a. The results were compared with heat pipes in the straight configuration. Fig. 6b depicts the thermal resistance of the bent heat pipes as a function of different heat power inputs. In general, the thermal resistance increment with increasing bending angles at lower heating powers (2 W, 4 W, 6 W) is distinguishable as the heat pipe effect was not effectively activated and thermal resistance of the pipe was high. As the heating power increases the thermal resistance difference gradually diminished. When the power was larger than 8 W, the thermal resistance curves were almost overlapped and the influence of bending was too subtle to be observed.

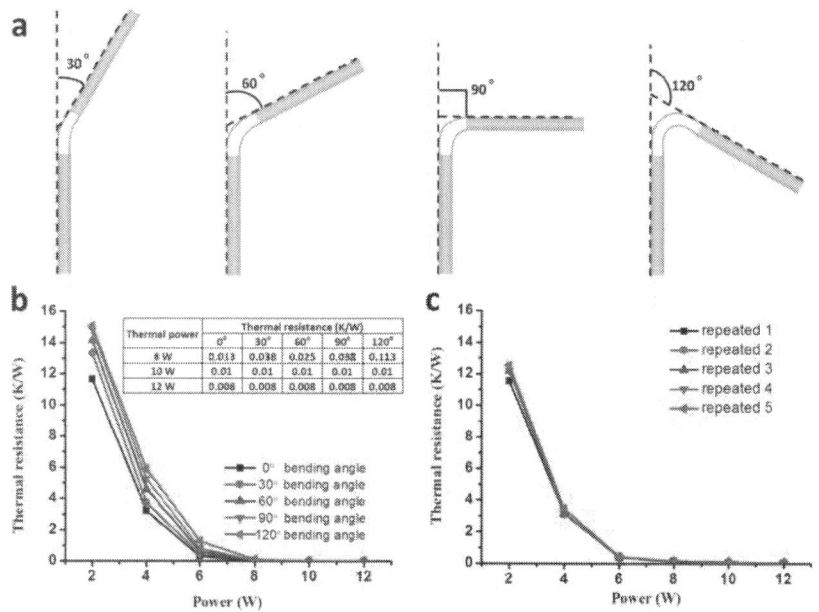

Figure 6. (a) Schematic illustration of bending experiments; (b) thermal resistance of flexible heat pipes filled with 30% working fluid bent at different angles; and (c) thermal resistance after repeated bending tests.

When the heat pipe bends, the cross-sectional area for the vapor channel would be reduced and the normal vapor flow in straight pipes would be interrupted. Specifically, in the curved pipes the secondary rotating flows, which were induced by the combined friction force at heat pipe walls and the centrifugal force, would add to the original main flow along the pipe axial. As summarized by Wongwises and Naphon [20], the correlation factor for vapor flow in curved tubes and straight tubes can be empirically described by:

$$\frac{f_c}{f_s} = 1 + f(R_e, D_e)$$

where f_c is the flow resistance factor of curved tubes, f_s is the flow resistance factor of straight tubes, the $f(R_e, D_e)$ term is a function of Reynolds number (R_e) and Dean number (D_e). This term is mainly related to the relative diameter of the cylindrical heat pipe and radius of curvature of the bent polyurethane tubes. The higher f_c causes additional pressure drop resulting in extra thermal resistances. In our case, the small ratio of the channel diameter to radius of curvature would only produce a small pressure drop in the channel flow. Furthermore, the strong mechanical robustness of the cylindrical polyurethane tubes allows for conformal bending instead of being squeezed, thus ensuring smooth vapor flow. These two factors would minimize the bending influence on the thermal resistance of the fabricated flexible heat pipes.

It is worthy pointing out that in our gravity-assisted operation mode bending also affects the gravitational contribution. When bending angle increases to 90° the upper half section of the heat pipe is oriented to the horizontal configuration where the axial gravitational pressure drop was absent. When further increasing the bending angle to 120°, gravity became an opposing force to the capillary pumping force. Indeed, we did observe higher thermal resistances for heat pipes bent at large angles (90° and 120°). But the thermal resistance difference was almost negligible under high power inputs at the evaporator section. High heating power induces high vapor pressure and weakens the influence of flow resistance on heat-transfer. The good heat-transfer performance of the flexible heat pipes should be attributed to their strong capillary pumping capabilities that were associated with the bioinspired superhydrophilic wick structures.

A preliminary reliability test of the fabricated heat pipes was conducted by repeated measurement of their thermal resistances at a fixed bending angle of 90°. Every heating test by increasing the power input from 2 W to 12 W lasts for more than 2 h. Fig. 6c shows that the fabricated heat pipes demonstrated consistent and stable thermal resistances and no obvious degradation of thermal performance was observed after more than 10 h operation. It should be mentioned that after the repeated cycling tests the original as-fabricated superhydrophilic $Cu(OH)_2$ wick layer was converted into thermodynamically more stable CuO at high temperatures. This transition was confirmed by XRD analysis of the wick meshes within the tested heat pipes. However, the converted hierarchical CuO wick meshes still preserve their excellent wettability with water. The unique wicking structure design and utilization of copper tubes in the evaporator and condenser achieved an overall low thermal resistance of the flexible heat pipe. This low thermal resistance enabled effective and timely heat transfer from evaporator to condenser, which in turn benefits the good reliability of fabricated flexible heat pipes. In the future work, metal laminated polymeric flexible connectors will be explored to realize higher thermal power delivery and thermal management of high-power electronic devices.

CONCLUSIONS

In this work, a bioinspired high-performance flexible heat pipe has been successfully developed by a simple and cost-effective low temperature process. Different from widely explored whole polymeric heat pipes, a flexible polyurethane tube was only used at the adiabatic section to connect copper tubes at the evaporator and condenser to minimize the overall thermal resistance and to mimic the flexibility of the heat conducting blood vessel in human body. Strong-base-oxidized superhydrophilic copper meshes bearing bioinspired hierarchical micro/nano-structures were utilized as the wicking material. Tested under a vertically gravity-assisted configuration, the results showed that bending had almost negligible influence on the thermal resistance of the fabricated heat pipes especially under high heating power inputs at the evaporator section. The combined flexibility and the consistent thermal performance could be attributed to the flexible polymeric connector design and the strong capillary pumping from the bioinspired superhydrophilic wicking structure. It is anticipated the higher power and reliable flexible heat pipes could be also developed by the similar approach, offering a powerful thermal management tool for flexible and wearable electronic devices.

ACKNOWLEDGMENT

This work was supported by the National Natural Science Foundation of China (NSFC, Grant nos: 51403127, 51420105009, 91333115 and 21401129), the Natural Science Foundation of Shanghai (Grant nos: 13ZR1421500 and 14ZR1423300), the Starting Foundation for New Teacher of Shanghai Jiao Tong University (No. 14X100040046), and the Zhi-Yuan Endowed fund from Shanghai Jiao Tong University.

REFERENCES

1. G.P. Peterson, An Introduction to Heat Pipes: Modeling, Testing, and Applications, John Wiley & Sons Inc., New York, 1994.
2. L.L. Vasiliev, Appl. Therm. Eng. 25 (2005) 1–19.
3. A. Faghri, J. Heat Transf. 134 (2012) 123001.
4. F.E. Bliss, E.G. Clark, B. Stein, in: Proceedings of ASME Space Systems and Thermal Technologies for the 70's, 1970, pp. 1–7.
5. B.R. Babin, G.P. Peterson, J. Heat Transf. 112 (1990) 602–607.
6. D. Harris, D. Odhekar, Front. Heat Pipe 2 (2011) 023002.
7. Y.X. Wang, G.P. Peterson, J. Thermophys. Heat Transf. 17 (2003) 354–359.
8. K. Tanaka, Y. Abe, M. Nakagawa, C. Piccolo, R. Savino, Ann. N. Y. Acad. Sci. 1161 (2009) 554–561.
9. C. Oshman, B. Shi, C. Li, R. Yang, Y.C. Lee, G.P. Peterson, V.M. Bright, J. Microelectromech. Syst. 20 (2011) 410–417.
10. G.-W. Wu, W.-P. Shih, S.-L. Chen, in: Proceedings of the 10th International Heat Pipe Symposium, 2011, pp. 80–85.
11. C. Oshman, Q. Li, L.A. Liew, R. Yang, V.M. Bright, Y.C. Lee, J. Micromech. Microeng. 23 (2013) 015001.
12. S. Ogata, E. Sukegawa, T. Kimura, in: Proceedings of the IEEE Intersociety Conference on Thermal and Thermomechanical Phenomena in Electronic Systems, 2014, pp. 519–526.
13. S.S. Hsieh, Y.R. Yang, Energy Convers. Manag. 70 (2013) 10–19.
14. Y. Ji, G. Liu, H. Ma, G. Li, Y. Sun, Appl. Therm. Eng. 61 (2013) 690–697.
15. D.D. Odhekar, D.K. Harris, in: Proceedings of the Tenth Intersociety Conference on Thermal and Thermomechanical Phenomena in Electronics Systems, 2006, pp. 570–577.
16. C. Li, G.P. Peterson, Int. J. Heat Mass Transf. 49 (2006) 4095–4105.
17. P. Tao, W. Shang, C. Song, Q. Shen, F. Zhang, Z. Luo, N. Yi, D. Zhang, T. Deng, Adv. Mater. 27 (2015) 428–463.
18. H. Hou, Y. Xie, Q. Li, Cryst. Growth Des. 5 (2005) 201–205.
19. S. Lips, F. Lefèvre, J. Bonjour, Int. J. Heat Mass Transf. 53 (2010) 694–702.
20. P. Naphon, S. Wongwises, Renew. Sustain. Energy Rev. 10 (2006) 463–490.

CITATION

Chao Yang, Chengyi Song, Wen Shang, Peng Tao, Tao Deng, Flexible heat pipes with integrated bioinspired design, Progress in Natural Science: Materials International, Volume 25, Issue 1, February 2015, Pages 51-57, ISSN 1002-0071, http://dx.doi.org/10.1016/j.pnsc.2015.01.011.

CHAPTER 7

Rehabilitation Priority Determination of Water Pipes Based on Hydraulic Importance

Do Guen Yoo [1], Doosun Kang [2], Hwandon Jun [3] and Joong Hoon Kim [4],

[1]Research Center for Disaster Prevention Science and Technology, Korea University, Seoul 136-713, Korea;
[2]Department of Civil Engineering, Kyung Hee University, 1732 Deogyeong-daero, Giheung-gu, Yongin-si 446-701, Gyeonggi-do, Korea
[3]Department of Civil Engineering, Seoul National University of Science and Technology, 172, Gongreung 2-dong, Nowon-Gu, Seoul 139-743, Korea
[4]School of Civil, Environmental and Architectural Engineering, Korea University, Anam-ro 145, Seongbuk-gu, Seoul 136-713, Korea

ABSTRACT

This paper describes a study conducted to develop a method to facilitate more reliable determination of the rehabilitation priority order for water pipes by taking into account the pipes' hydraulic importance. Existing methods use only the pipeline deterioration rate to determine the rehabilitation priority order. Accordingly, the deterioration rate under normal conditions and the hydraulic importance under abnormal conditions of water distribution pipelines were classified according to two different attributes. The deterioration rate of a water distribution pipeline was calculated in terms of the deterioration rate due to pipeline information factors and the deterioration rate resulting from the installation environment/external factors. The hydraulic importance of water distribution pipelines was calculated by considering the importance of a single pipe failure caused by water leakage or an accident and that of a multiple pipe failure caused by a disaster, such as an earthquake. These four attribute factors were employed in a multi-criteria decision-making process called a weighted utopian approach, developed in this study that determines the final rehabilitation priority order for each pipeline. The study results indicate that the rehabilitation priority order can be determined more easily using this approach than with previously-developed methods and that the model developed is easier and more convenient to apply than existing rehabilitation priority order models that require a large amount of data, as well as complex failure probabilities and mathematical models.

INTRODUCTION

A water distribution system is an important part of the social infrastructure, facilitating water transport, distribution and supply. Such a system is a highly complicated network that combines pipelines, pumps and valves. Hence, the facilities in any such system should be continuously improved and updated based on specific plans to maintain the stability and safety of the water supply. As the importance of maintaining and managing this water distribution system has increased, projects for repairing and replacing deteriorating water pipes are currently being undertaken throughout Korea. However, the current methods used to deal with deteriorating pipelines involve an evaluation of the degree of deterioration based on empirical means, as well as reactive rehabilitation projects undertaken after accidents, leading to economic losses and failure to improve system functions. The current approach to determining the rehabilitation priority order for pipelines is based only on the year of installation of the pipes, with no clear criteria for evaluating the degree of deterioration. To address these problems, a new approach to determine the rehabilitation priority order for a water distribution system should be developed that overcomes the drawbacks of the existing methods. In addition, the rehabilitation priority order should be determined according to not only the physical deterioration rate of individual pipes, but also the relative importance of those pipes, to increase the overall stability and safety of the system.

Early studies of the methods to determine the rehabilitation priority order for water distribution systems were conducted using rehabilitation models based on empirical determination that use only general guidelines. Subsequent studies on the determination of the rehabilitation priority order can be broadly classified into the categories of analysis of the deterioration rate and failure probability, regression analysis coupled with failure probability analysis and priority order estimation based on economic feasibility analysis. The Guidance Manual-Water Mains Evaluation for Rehabilitation/Replacement published by the American Water Works Association Research Foundation [1] proposes physical strength, water quality in the pipes, hydraulic conditions and water leakage as criteria for evaluating pipeline functionality. Accordingly, the manual notes that comprehensive management that considers these four criteria is required to effectively maintain and manage water distribution pipelines. The Water

INTRODUCTION

Research Centre [2] of the United Kingdom states that it is reasonable to prioritize the repair and rehabilitation of water distribution pipes that are frequently associated with accidents. K-water [3] determined pipeline deterioration rates and developed a weighting system for pipeline repair and rehabilitation by collecting and summarizing domestic and international literature, as well as data on large-diameter water distribution pipes in metropolitan water distribution systems. Subsequently, K-water developed a model to estimate the deterioration rate of water pipes using a point-based evaluation method that considers the estimated pipe condition and weight, as well as a model for prioritizing the replacement and rehabilitation of water pipes. However, this model is limited in that the point-based evaluation method does not reflect the hydraulic characteristics of water distribution systems. Kim et al. [4] re-calculated the cost function by modifying and complementing the failure rate function proposed by Shamir and Howard [5]. In addition, they estimated the deterioration rate using a probabilistic neural network (PNN) and proposed a rehabilitation model that prioritizes the rehabilitation and replacement of pipes using a shortest-path model. Studies in which regression analysis and failure probability analysis have been applied to estimating the rate of deterioration of water pipes have been conducted by Marks and Clark [6,7], Agbenowosi [8] and Park and Loganathan [9,10]. Marks and Clark [6,7] proposed a method for failure modeling based on the deterioration rate of the water distribution pipes. After classifying the deterioration status of water distribution pipes into "early stage", with a small number of failures, or "later stage", with a large number of failures, their method applies these stages to each case. Based on their results, they proposed that the deterioration status of a water distribution pipe could be represented by a proportional hazards model in the early stage and a Poisson-type model in the later stage. Deb et al. [11] classified water distribution pipes according to their installation year, pipe material, and diameter and backfilled soil type. They also developed a probabilistic model called "KANEW" based on this classification to estimate the number of pipes in a water distribution system that should be replaced annually. Agbenowosi [8] represented the factors related to water distribution pipes in a pipe load model and a pipe break model, so that the factors could be analyzed mathematically and the most economical time at which to replace a water distribution pipe could be determined. Park and Loganathan [9,10] proposed a failure estimation model for water distribution pipes using a threshold break rate. Using this model, the economically optimal replacement time is determined as a function of the costs of pipe replacement and rehabilitation, a discount rate and the length and diameter of the pipes.

In general, these models, which determine the rehabilitation priority order of pipes in water distribution systems using regression and failure probability analyses, predict pipe failure through complex formulas and determine the most economical replacement and rehabilitation times through economic feasibility analysis. While these models can predict pipe failure through failure probability analysis, they require a large amount of basic data and complex formulas to produce results. Models for determining the rehabilitation priority order of pipes in water distribution systems on the basis of economic feasibility, which have been examined by Shamir and Howard [5], Walski [12] and Luong and Fujiwara [13], achieve maximum efficiency at minimum cost. Alvisi and Franchini [14] proposed a near-optimal rehabilitation scheduling method based on a multi-objective genetic algorithm. With reference to a fixed time horizon, the goal is to minimize the overall costs of repairing and/or replacing pipes and to maximize the hydraulic performances of the water network. However, given that these models also require a large amount of basic data and that water distribution pipes are part of social infrastructure networks that directly affect public welfare, cost optimization cannot be an essential condition.

In recent years, computer-aided models and decision support tools, such as Care-W [15, 16, 17, 18, 19, 20, 21], CASSES [22] and AWARE-P [23], have been developed. The CARE-W project aimed to develop methods and software that would enable engineers of the water undertaking to define and implement an effective management of their water supply networks, rehabilitating the right pipelines at the right time. This project was organized in eight work packages (WP), which were the construction of a control panel of performance indicators (WP1), description and validation of technical tools (WP2), elaboration of a decision support system for annual rehabilitation programs (WP3), elaboration of long-term strategic planning and investment (WP4), elaboration of the CARE-W prototype (WP5), testing and validation of the CARE-W prototype (WP6), dissemination (WP7) and project management (WP8). CARE-W used four probabilistic forecast failure models (proportional hazard model, Markov model, Poisson analysis, and non-homogeneous Poisson process) and three mathematical hydraulic reliability assessment models (Aquarel, Failnet-Reliab, Relnet). Among these WPs, WP3 and WP4 focused on the determination of priority and scheduling for rehabilitation according to the target time period (one year and long-term plan). They used multi-criteria techniques, such as scoring and ELETRE TRI methods, to determine relevant procedures during annual rehabilitation programs. In case of long-term rehabilitation strategies, extended KANEW is used. In the continuation of CARE-W, a new break prediction model, called linear extension of the Yule process (LEYP), was developed. It involved a statistical model based on a counting process and relied not only on the pipe's characteristics and environment, but also its age and previous breaks. This new model was chosen to be used in the development of the break prediction software, CASSES. The main result from this software was the number of breaks for each pipe for a period in the future. The AWARE-P project has been performed for providing water and wastewater utilities with the know-how and the tools needed for efficient decision-making in infrastructural asset management (IAM) of urban water services. The key point of this project is IAM as a management process, based on plan-do-check-act (PDCA) principles and requiring alignments between the organization's strategic objectives and targets and the actual priorities and actions implemented.

Water distribution pipes can be divided into hydraulically significant or insignificant pipes depending on the installation location, valve location, base demand for water at nodes, variation in demand, pipe flow rates and the population served. Accordingly, the importance of individual water distribution pipes can be termed "hydraulic importance." This study was conducted to develop a method to determine the rehabilitation priority order of pipes in water distribution systems on the basis of both the deterioration rate of water pipes and their hydraulic importance. Using the method proposed in this paper, pipes are first classified according to their deterioration rate and hydraulic importance according to two different attributes. A multi-criteria decision-making method called a weighted utopian approach, which combines a weighting method and a distance measurement method, is then used to prioritize the pipes' rehabilitation needs.

METHODOLOGIES

In this study, the deterioration rate under a normal situation and hydraulic importance under abnormal conditions of water distribution pipes were used to determine the rehabilitation priority order of pipes in a water distribution system, as shown in Figure 1. The deterioration rate was divided into two components: the deterioration rate due to internal factors and the deterioration rate due to installation environment factors (external factors). The hydraulic importance was also divided into two components: the importance related to single pipe failure (which accounts for most pipe accidents) and the importance related to multiple pipe failures due to disasters, such as earthquakes.

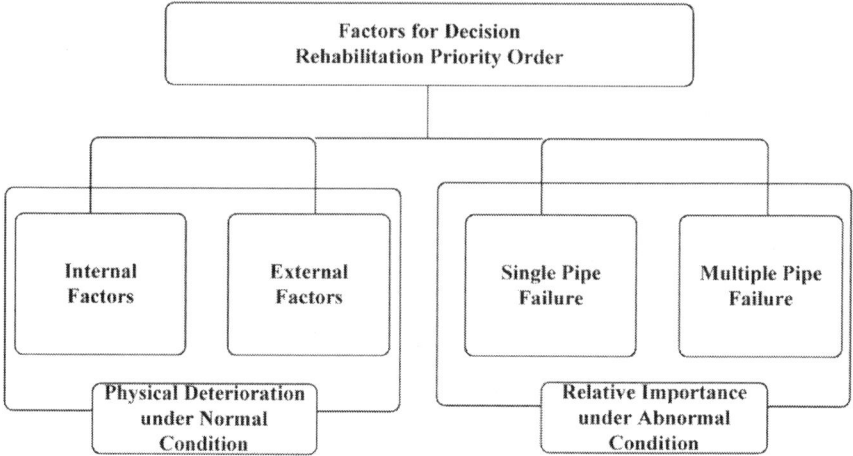

Figure 1 Attributes for rehabilitation priority order.

Pipe Deterioration

In this study, the deterioration rate calculation method and factors proposed by Kim *et al.* [4] were considered, and a total of 12 factors were classified as either internal or external factors. Eight internal factors were identified: pipe material, pipe diameter, interior corrosion rate, elapsed time since installation, type of joint, record of leakage and breakage, record of civil appeals regarding water quality and pressure and maximum pipe pressure. Four external factors were identified: exterior corrosion rate, backfilled soil type, road width and installation district. To quantify the deterioration rate, weights for the factors should be taken into consideration. From among the methods available for determining such weights, an eigenvector method employed for weight calculation in an analytic hierarchy process (AHP) [24], a decision-making method, was selected for use in this study.

A PNN model proposed by Specht [25] was employed to take the weight of each factor into consideration in calculating the degree of deterioration. To apply this PNN model to the prioritization method developed in this study, the conditions of the factors should be classified according to conditional values. For example, the factors that are most heavily influenced by the deterioration rate are assigned a value of 1.0, and those that are influenced the least are assigned a value of 0.0. The conditions that have the greatest influence on a factor are assigned a value of 1.0, and those with the least influence are assigned a factor of 0.0. The conditions of factors, such as the exterior corrosion rate, backfilled soil type, road width and installation district, which are used to calculate the possibility of water distribution pipes failing due to installation environment/external factors, were classified according to a total of five conditional values. The relative weight of each factor is then calculated using the eigenvector method. The PNN model applied in this study consists of input, pattern, summation and output layers, as illustrated inFigure 2. In the input layer, a conditional value defined according to conditions for standard pipes is assigned to each neuron, while five large neurons are placed in the pattern layer, with one large neuron being responsible for a certain boundary condition value. A large neuron is one of the main components when we make the structure of a PNN. In this study, a large neuron works as a dividing criterion, which is divided into five deterioration groups among the pipes. That is, the first large neuron includes neurons that have a conditional value of 1, while the second, third, fourth, and fifth large neurons consist of small learning neurons with conditional values of 0.75, 0.5, 0.25 and 0.0, respectively. The four small learning neurons correspond to the exterior corrosion rate, backfilled soil type, road width and installation district, respectively. The pattern layer plays a critical role in identifying input data and performing classification, by measuring the distance between the input data and learning neurons, and includes the calculated distance in the activation function. In the model used in this study, a Gaussian function was used to determine the width and area of the data in the activation function. The summation layer has a single summation neuron that simply adds data learned from the pattern layer. The output layer has five neurons that have a simple classification function. When classification is performed in the output layer, classification is not done based on a percentage value, but by a simple summation of all values obtained in the summation layer, once trained, after the input data are entered. When output values are defined as $P1$, $P2$, $P3$, $P4$ and $P5$; $P1$ means a probability value that is to be included in the neuron responsible for the conditional value of 1. From among $P1$, $P2$, $P3$, $P4$ and $P5$, obtained through the process above, the largest value is selected; if $P1$ is the largest value, the deterioration rate of the corresponding pipe is classified into Group 1. For example, if $P1$, $P2$, $P3$, $P4$ and $P5$ have values of 0.45, 0.15, 0.2, 0.15 and 0.05, respectively, as a result of calculating the deterioration rate, $P1$, which has the largest probability, is selected, and the deterioration rate of the corresponding pipe becomes that corresponding to Group 1. The groups and probability values of the pipes are obtained by using the PNN algorithm to assess the deterioration rate of the pipes based on internal factors and installation environment/external factors. In addition, values are distributed between 0 and 1 using an interpolation method to compare the values for each pipe.

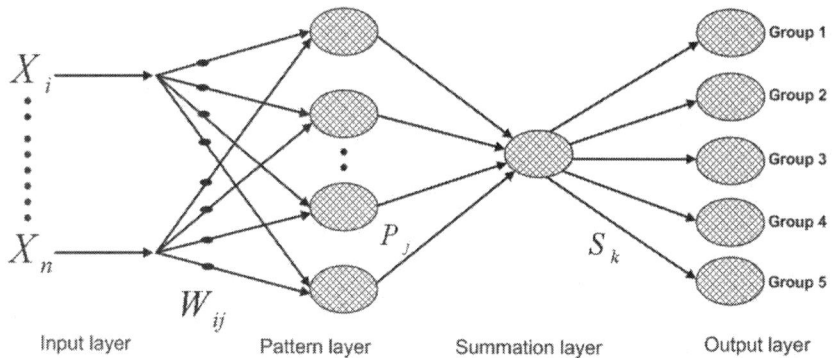

Figure 2 Probabilistic neural networks for assessing pipe deterioration.

Hydraulic Importance: Single Pipe Failure

Whereas most existing methods restrict the range of damage due to the failure of a water pipe to the corresponding water pipes, Jun *et al.* [26] introduced the concept of "unintended isolation," in addition to the concept of partial segments in water distribution networks, as proposed by Walski [27], which are used to estimate the damage area due to the failure of water pipes. According to Walski [27], a segment can be defined as a set of surrounding pipes that are closed by water control valves, along with a failed pipe, in the event of a failure. That is, when a failed pipe and the surrounding pipes should be isolated, a segment can be defined as the number of separated water pipes and nodes. When the area of damage is estimated using only failed water pipes, it is limited to those cases in which only the failed water pipes are closed. However, considering the practicalities, adjacent pipes may also have to be closed, depending on the number and locations of the water control valves. For example, Figure 3 shows that when P2 fails, such that the water control valves have to be closed to enable repairs, P4 is also closed. Consequently, the water supply to node N1 within the closed segment is also stopped. This damage area estimation method can produce more realistic estimates of the areas of damage, because it considers the number and locations of the water control valves. When a segment that includes a failed water pipe is closed to enable repair, the water supply to other sub-pipes connected to the failed pipes is also shut off if the segment constitutes the only path from the water source. That is, some pipes can experience disconnection of the water supply due to the segment closure. Jun *et al.* [26] defined this event as unintended isolation and developed an updated breadth-first search algorithm based on a node-arc matrix, which is more advanced than a conventional breadth-first search algorithm, to identify the unintentionally isolated pipes. In the event of unintended isolation, consumers of water in the area are deprived of their supply until the failed water pipe is fixed. That is, even if the supply in the area is not shut down by the water control valves, consumers in the area are isolated from the water supply being provided to those living in the segment. In Figure 3, a segment consisting of P6 and N6 generates unintended isolation.

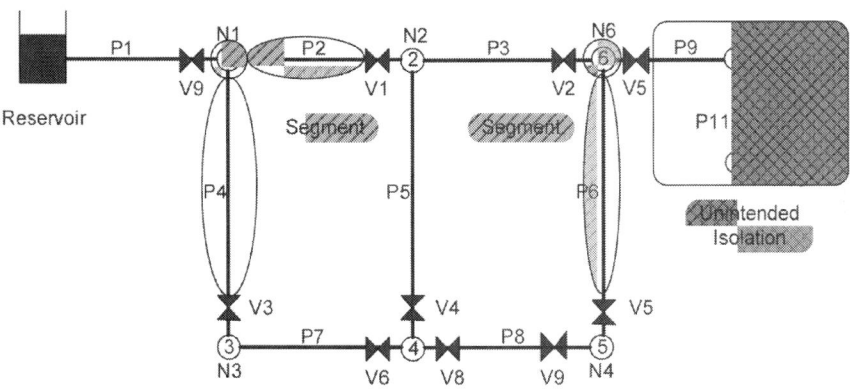

Figure 3. Segment and unintended isolation [26]. P, probability value; N, node.

To determine the importance of individual water pipes in the event of a single pipe failure, it is necessary to assume the following: the failure probability of each pipe is the same, and no more than two pipes can fail at any one time. The probability of pipe failure can differ among pipes according to the deterioration rate of the pipe, size, material, *etc.* In the case of hydraulic importance calculations of single pipe failure simulation, components of failures include not only unintentional failure (e.g., breakage and leakage), but also intentional failure (e.g., cut off the water). In this case, we cannot exactly predict intentional failure. Because of that, we assumed that the failure probability of each pipe is the same. It is also assumed that all of the valves are functioning properly, so valve malfunctions do not have to be considered. Based on these assumptions, the segments and instances of unintended isolation that are generated as a result of a single pipe failure in a water distribution system are calculated, and this calculation process is executed for all pipes. Finally, the importance of an individual water pipe in the event of a single pipe failure is calculated for each of the water pipes, based on Equation (1). The greater the $HISPF_i$ (hydraulic importance by single pipe failure when pipe i fails) value of an individual pipe, the higher the relative importance of the pipe in the event of a failure. Conversely, the smaller the $HISPF_i$ value, the lower the relative importance of the pipe in the event of a failure.

$$\text{HISPF}_i = \frac{Q_{i,S} + Q_{i,UI}}{Q} \tag{1}$$

where HISPF_i = hydraulic importance by single pipe failure when pipe i fails; Q = total pipe flow under normal conditions; $Q_{i,S}$ = segment pipe flow when pipe i has failed; $Q_{i,UI}$ = unintended isolation pipe flow when pipe i has failed.

Hydraulic Importance: Multiple Pipe Failures

Because earthquakes typically lead to major damage and multiple pipe failures, the importance of water pipes as evaluated in the event of an earthquake was considered in this study. Yoo *et al.* [28,29] developed the reliability evaluation model of seismic hazards for water supply networks (REVAS.NET) for use in evaluating seismic risks to water distribution systems. This model is intended to prevent damage from earthquakes by calculating the reliability of a water distribution system in the event of an earthquake. This model was employed in this study. The process by which REVAS.NET was employed and the reliability factors used in this study are as follows:

Step (1) Establishment of basic topological structure of water supply network in the area of application;
Step (2) Construction of fragility curves for each component of a water distribution system;
Step (3) Simulated earthquake generation (seismic location and magnitude);
Step (4) Seismic wave attenuation;
Step (5) Determination of component conditions;
Step (6) Execution of EPANET hydraulic analysis;
Step (7) Negative pressure processing;
Step (8) Calculation of hydraulic reliability of systems/components.

Once the states of the water pipelines, tanks and pump facilities after the simulated earthquake have been determined, these states can be simulated via hydraulic analysis, and the hydraulic analysis results for the water distribution system can be derived. To quantify the results, a reliability factor is used. In general, system reliability, which is a widely-used concept in many fields, refers to the probability of a service running continuously within a system. In REVAS.NET, nodal serviceability (N_S) and system serviceability (S_S) [30,31,32], which are known to be relevant to a water distribution system, are used as reliability factors.

In this study, the hydraulic importance of water pipelines was calculated using the nodal serviceability (N_S), expressed by Equation (2). In the case of N_S, the hydraulic importance of a node for which nodal demand is required is calculated from the ratio of the available nodal demand to the required nodal demand. The usability of a node at which the required nodal demand is not present is assessed by evaluating the ratio of the pressure after the earthquake to the allowable minimum nodal pressure. This reliability index, which represents the serviceability that each node can provide, is calculated from the required nodal demand and minimum pressure, so it can be used to assess the relative supply ability per node.

$$\text{Nodal Serviceability } (N_{S,i}) = \begin{cases} \dfrac{Q_{avl,i}}{Q_{in\,i,i}} & \text{when } Q_{in\,i,i} \neq 0 \\ \sqrt{\dfrac{\text{Min}(P_i, P_{min})}{P_{min}}} & \text{when } Q_{in\,i,i} = 0 \end{cases}$$

(2)

where $Q_{avl,i}$ = available nodal demand at node i; $Q_{ini,i}$ = required nodal demand at node i; P_i = nodal pressure at node i; P_{min} = allowable minimum nodal pressure.

Finally, the importance of a water pipeline in the event of multiple pipe failures is calculated using Equation (3). The importance can be calculated by subtracting the average reliability value of each node connected to each water pipe from value 1. The higher the resulting value is, the lower the serviceability of the water pipelines related to the corresponding water pipeline as a result of a corresponding pipeline breakage followed by multiple pipe failures.

$$\text{HIMPF}_i = 1 - \frac{UN_{S,i} + DN_{S,i}}{2}$$

(3)

where HISPF_i = hydraulic importance of pipe i by multi-pipe failure; $UN_{S,i}$ = upstream nodal serviceability of pipe i; $DN_{S,i}$ = downstream nodal serviceability of pipe i.

Weights of Attributes

Many studies have been conducted to examine methods for determining the weighting factors of attributes used in multi-criteria decision-making methods. These studies have shown that common usage standards do not exist among the conventional methods and that no one method stands out as better than the others. Although some studies have proposed weighting factor methods based on the type of decision-making domain, the theoretical validity of these methods has not been confirmed. Thus, it is crucial to determine appropriate weighting methods on the basis of the types and attributes of the decision-making domain and to make a reasonable decision accordingly. In this study, the five weighting methods listed in Table 1 were employed.

Table 1. Weighting methods.

No.	References	Method
1	-	Same weighting
2	[24]	Eigenvector
3	[33]	Churchman–Ackoff
4	[34]	Rating
5	[35,36]	Entropic

Eigenvector

The eigenvector method proposed by Saaty [24] has been used as a weighting calculation method for AHP, which is one of the main multi-criteria decision methods. Weighting factor values can be calculated using Equation (4) below. This is a widely used method, because it is easily applied and suitable for problem analysis.

$$A\varpi = \lambda_{max} \varpi$$

(4)

where A = positive pairwise comparison reciprocal matrix; λ_{max} = maximum eigenvalue of A; ϖ = weighting factor of the attribute.

Churchman–Ackoff Method

The Churchman–Ackoff method [33] is based on rankings, whereby weightings are calculated via Equation (5). For example, when a problem consists of five attributes, n is 5 and ¢j=15j is 15. When Attributes 1 to 5 are ranked from 1st to 5th, respectively, the weights of the attributes will be 0.33, 0.27, 0.20, 0.13 and 0.07, in sequence.

$$\omega_i = \cfrac{J}{\sum_{j=1}^{n} j} = \cfrac{2J}{n(n+1)} \qquad (5)$$

where j = attribute (j=1,2,...,n); n = number of attributes; J = rank of the importance of the j-th attribute.

Rating

The rating method [34] calculates weights based on ratings received from a respondent, and calculation is performed as indicated in Equation (6). The scale of the rating can be arbitrarily established from 0 to 10.

$$\omega_i = \cfrac{\sum_{k=1}^{l} \omega_{jk}}{\sum_{j=1}^{n}\sum_{k=1}^{l} \omega_{jk}} \qquad (6)$$

where j = attribute (j=1, 2,..., n); n = number of attributes; k = respondent (k=1, 2,..., l); l = number of respondents; $\omega_{jk} = \cfrac{r_{jk}}{\sum_{j=1}^{n} r_{jk}}$; r_{jk} = rating of respondent k.

Entropic Method

The entropic method was proposed by Hwang and Yoon [35] and Soofi [36]. It is an objective method that is used to calculate weights using only data in the decision-making domain and not using the subjective judgments of decision makers. The procedure for the entropic method starts by estimating the weight vectors from an alternative-attribute matrix. From the viewpoint of entropy, an alternative-attribute matrix contains information that can be used to estimate weights for criteria. That is, a criterion that has a large difference between alternatives is an important criterion, whereas a criterion that has a small difference between alternatives is a less important criterion. It is assumed that a decision-making domain can be represented by matrix D, shown in Equation (7) below.

$$D = \begin{bmatrix} x_{11} & \cdots & x_{1j} & \cdots & x_{1n} \\ \vdots & \cdots & \vdots & \cdots & \vdots \\ x_{i1} & \cdots & x_{ij} & \cdots & x_{in} \\ \vdots & \cdots & \vdots & \cdots & \vdots \\ x_{m1} & \cdots & x_{mj} & \cdots & x_{mn} \end{bmatrix}$$

(7)

If pij is a normalized result with respect to all attributes, pij can be represented by Equation (8).

$$P_{ij} = \frac{x_{ij}}{\sum_{i=1}^{m} x_{ij}} \quad (i=1,2,\ldots,m; j=1,2,\ldots,n)$$

(8)

Assuming that the entropy of each attribute is Ej, the entropy can be calculated using Equation (9).

$$E_j = -k \sum_{i=1}^{m} P_{ij} \cdot \log P_{ij} \quad \left(k = \frac{1}{\log m}\right)$$

(9)

To calculate the weight of an attribute, a level of diversity, dj, is used that can be calculated using Equation (10). When these values are normalized with respect to every attribute, this represents the weight of the corresponding attribute (Equation (11)).

$$d_j = 1 - E_j$$

(10)

$$\omega_i = \frac{d_j}{\sum_{j=1}^{n} d_j}$$

(11)

Weighted Utopian Approach

The utopian approach proposed by Xanthopulos *et al.* [37] can be applied whenever there are multiple attributes and offers the advantage of convenience in application. However, it suffers the drawback that calculations are performed assuming the same weights for each attribute. In general, the weights of the attributes cannot be the same during decision making. If the same weights were to be used, it would produce results and judgments that differed from those of the decision makers.

Yoo *et al.* [38] proposed a weighted utopian approach (WUA) that combines a weighting method for multi-criteria decision making. A schematic view of the WUA is shown in Figure 4. Assuming that there are two attributes that enter into making a decision, namely A1 and A2, a coordinate plane with two axes, can be visualized. The normalization of the attribute values of every alternative between 0 and 1 over the coordinate plane can be seen in the graph on the left-hand side of Figure 4.

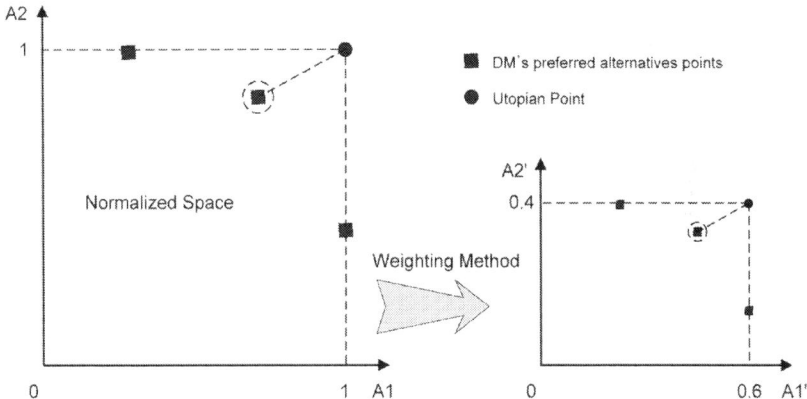

Figure 4. Weighted utopian approach.

Once the weights of the attributes have been calculated using the weighting method for multi-criteria decision making, they are multiplied for each axis, so that alternative attribute and utopian point values, distributed between 0 and 1, are redistributed on a coordinate plane. For example, if the weights of A1 and A2, which were calculated using the weighting method, are 0.6 and 0.4, respectively, the newly created utopian point on a graph changes from (1, 1) to (0.6, 0.4). Finally, using the redistributed graph, the distance between the utopian point and the alternative is calculated. The distance measurement method used to calculate a distance employs the commonly used Euclidean distance, illustrated in Figure 5. Finally, the priority of the alternatives is determined from the values of the Euclidean distance, starting with the shortest distance.

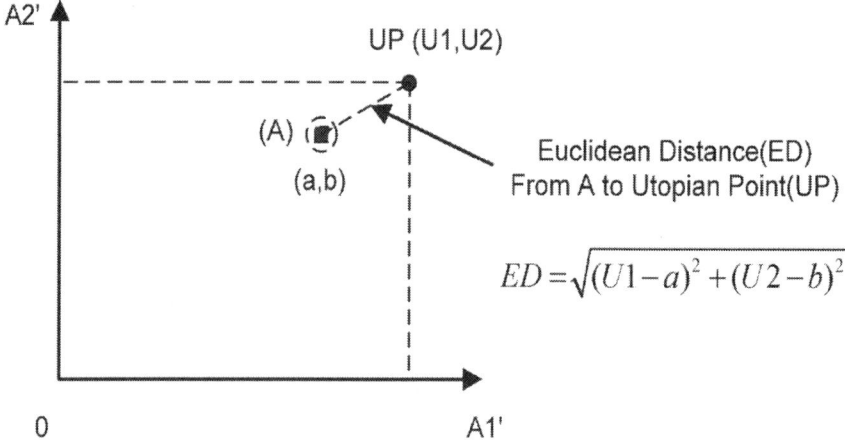

Figure 5 Euclidean distance.

That is, the priority of each alternative is determined using a weighting method, and the final priority ranking of the alternatives is determined by sorting the average ranking values produced using the weighting method in an ascending order.

Figure 6 shows a flow chart for the model proposed in this paper for determining the rehabilitation priority order for water pipelines.

APPLICATION RESULTS 169

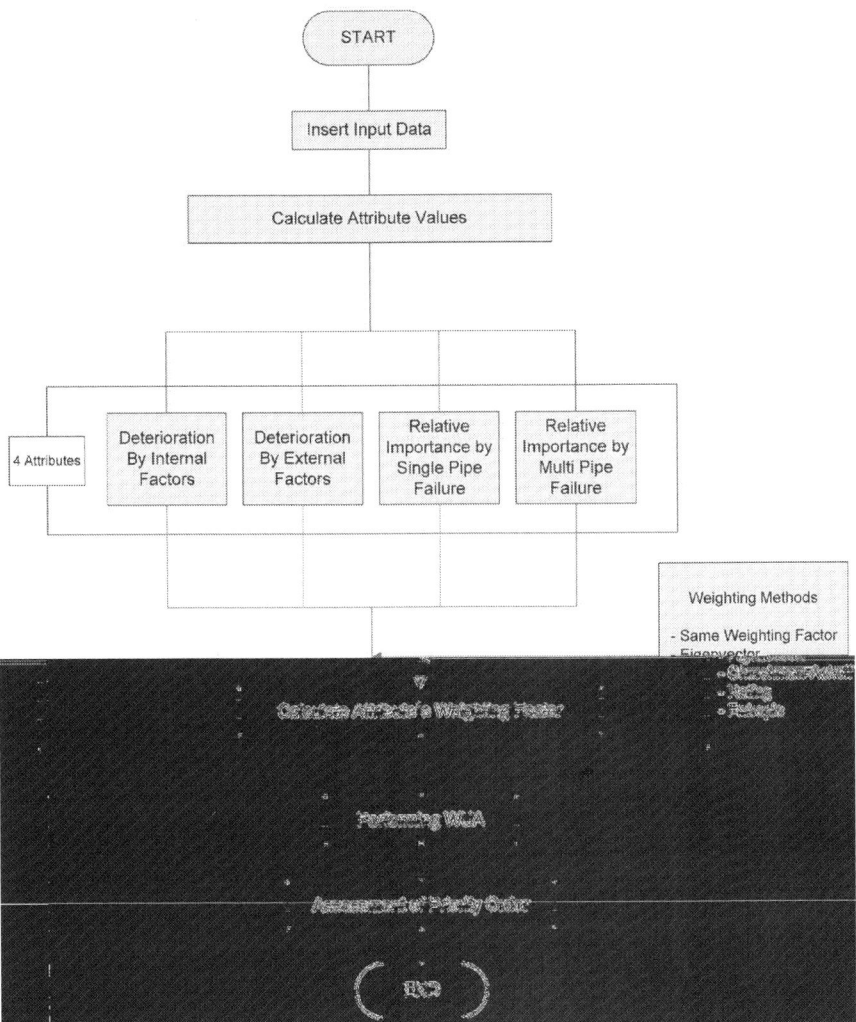

Figure 6 Flow chart for determination of rehabilitation order WUA, weighted utopian approach.

APPLICATION RESULTS

The model proposed in this study was applied to the KA network in City J, which divides its overall water network to enable area isolation, with 10 large district meter areas and 130 small district meter areas. City J has established main project plans to replace and rehabilitate deteriorating water pipelines in 50 small district meter areas and to maintain the water distribution facilities. There are 350 pipelines in the KA network, which is one of the 10 large district meter areas, with a total pipeline length of 35,951 km. The average demand per day is 16,058.4 cubic meters per day (CMD). The number of nodes is 230, and the number of installed valves is 160. Figure 7 shows the locations of pipelines and nodes in the KA network. Within the network, 24 pipes have diameters greater than 300 mm, while most of the pipes have diameters between 80 and 300 mm.

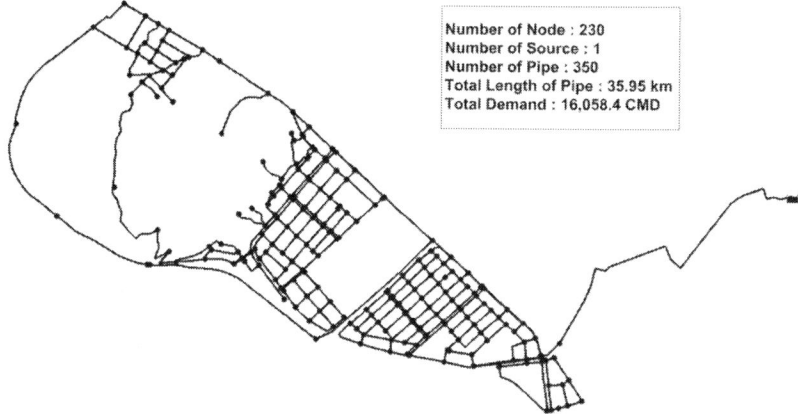

Figure 7 KA network in City J. CMD, cubic meters per day.

Results: Deterioration Rate

Input data consisting of internal factors and external factors for each pipe were acquired to calculate the deterioration rates of the water pipelines using the PNN algorithm. Scores for each factor, to be applied in the PNN algorithm, were determined according to Table 2 and Table 3. In addition, the weights for each factor were obtained by applying the eigenvector method to the results of surveys of 15 experts (the results are summarized in Table 4). Of the 15 experts, five were from academia, five were from corporations and five were from research centers and government agencies. The estimates of the deterioration rates with respect to the internal factors indicated that the number of Group 1 water pipes was 250 of the 350 pipes and that the number of Group 5 water pipes was 100. The reason for such a distinctive difference can be discerned from the years of installation of the water pipes. The average number of years that had elapsed from the installation date of the Group 1 water pipes was 18.5 years, which is considerably longer than that of the Group 5 water pipes, which was four years, many of the Group 5 water pipes having recently been replaced or rehabilitated. This means that those water pipes for which

APPLICATION RESULTS

many years had elapsed since the installation date had a greater inside corrosion rate and more frequent records of civil appeals and repairs than more recently installed water pipes. The classification of deterioration rates by installation environment/external factors placed 10 pipes in Group 2, 81 pipes in Group 3 and 259 pipes in Group 5. The main reason that most pipes were in Group 5 was that no water pipes with a significantly large exterior corrosion rate (a factor to which a large weight applied) existed. Another reason was that most of the water pipes were installed in sandy soil, which tends to inhibit deterioration. As a result, the deterioration rate associated with the installation environment/external factors was not high.

Table 2 Scores resulting from internal factors

Factor	Detailed Classification	Scaled Score	Factor	Detailed Classification	Scaled Score
Pipe material	CI *, CIP *	1.00	Pipe diameter (mm)	Below 80	1.00
	PE *, PVC *	0.75		80–100	0.75
	SP *	0.50		100–150	0.50
	HI-3P *	0.25		150–250	0.25
	DCIP *	0.00		Above 250	0.00
Inside corrosion rate (mm)	Below −2	1.00	Installation year (Installation duration, y)	Above 20	1.00
	−2–0.5	0.75		15–20	0.75
	−0.5–0.3	0.50		10–15	0.50
	−0.3–0	0.25		5–10	0.25
	Above 0	0.00		Below 5	0.00
Record of leakage or valve replacement (No. of cases/(y·m))	Above 10	1.00	Maximum pressure of pipes (kg/cm^2)	Above 5	1.00
	7–10	0.75		4–5	0.75
	3–7	0.50		3–4	0.50
	0–3	0.25		2–3	0.25
	0	0.00		Below 2	0.00
Joint type	Welding	1.00	Record of a civil appeal (water quality or pressure)	Yes	1.00
	Rubber ring	0.50		No	0.00
	Mechanic	0.00			

Notes: * CI/CIP: Cast Iron Pipe; PE: Poly-Ethylene Pipe; PVC: Poly-Vinyl Chloride Pipe; SP: Steel Pipe; HI-3P: High Impact (3-Layer) Pipe; DCIP: Ductile Cast Iron Pipe.

Table 3 Scores resulting from external factors

Factor	Detailed Classification	Scaled Score	Factor	Detailed Classification	Scaled Score
Outside corrosion rate (mm)	More than 9	1.00	Road width	Industrial road	1.00
	6–9	0.75		4-lane road	0.75
	4–6	0.50		2-land road	0.50
	2–4	0.25		Street	0.25
	Less than 2	0.00		Sidewalk and bare ground	0.00
Installation district	Factory	1.00	Refilled soil type	Clay	1.00
	Roadside	0.75		Silt	0.50
	Commercial district	0.50			
	Apartment	0.25		Sand	0.00
	Residential area, farmland	0.00			

Table 4 Weighting values resulting from internal and external Factors

Factor	Weight
Pipe material	0.452
Pipe diameter	0.286
Inside corrosion rate	1.864
Installation year	1.252
Joint type	0.515
Record of leakage or valve replacement	1.848
Record of a civil appeal	1.140
Maximum pressure of pipes	0.643
Outside corrosion rate	2.015
Refilled soil type	0.680
Road width	0.610
Installation district	0.694

Results: Hydraulic Importance of Single Pipe Failure

The number of segments, as determined by the locations of valves, calculated to estimate the importance of each water pipe in the event of a single pipe failure was found to be 50. The average calculated importance of a water pipe when each segment failed was found to be 0.083. This result indicates that the KA network has a low importance value in the event of a single pipe failure. This low importance value is attributable to the proper placement of the valves. However, several major segments had considerably higher levels of importance. For example, if a segment consisting of water pipes 975, 1104 and 1105 were to fail, unintended isolation would occur for as many as 174 water pipes. As a result, the importance of the corresponding pipes was calculated to be 0.6340, which is extremely high. As such, a high relative importance, such as that calculated for the above example, was generated in a total of four cases, except for those water pipes for which the water is received directly from the water source. These four segments are described in Table 5 and Figure 8. Water pipes that belong to such segments can present a major risk when breakage or emergency situations occur. Accordingly, they must always be carefully maintained and monitored. However, the importance of the other segments, which had a small number of water pipes and which did not give rise to unintended isolation (UI), was low.

APPLICATION RESULTS

Table 5 Value of importance of segment in the event of single pipe failure UI, unintended isolation.

Segment	Pipe ID	No. of Pipes by UI	Importance
1	975, 1104, 1105	174	0.6340
2	976	174	0.5868
3	1005, 1118	98	0.4582
4	1059	184	0.4156

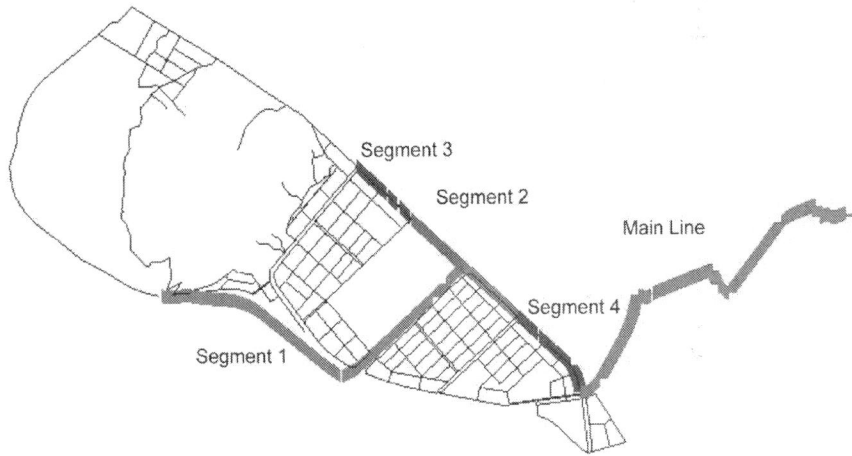

Figure 8. Importance of segments by single pipe failure.

Results: Hydraulic Importance of Multiple Pipe Failures

The REVAS.NET model was used in this study to calculate the importance of water pipes in the event of multiple pipe failures. As shown in Table 6, a previous earthquake of magnitude M7 that had occurred in Province J, which includes the KA network, was considered. The execution number for the Monte Carlo simulation was set to 10,000, while the minimum pressure in the KA network was considered to be 15 m (KWWA [39]).

Table 6. Reliability evaluation scenario and results for the KA network.

Seismic Hazard		Nodal Serviceability (Average Value) (N_S)
Historical Location Data (Number of Data Points)	Magnitude	
J Province (29)	Specific Magnitude (M = 7)	0.662

The results obtained with REVAS.NET indicate that the average system serviceability (S_S) and nodal serviceability (N_S) were 0.533 and 0.662, respectively. Figure 9 depicts the spatial distribution of the nodal serviceability and shows that the reliability was relatively low at the nodes of the terminal area, which are far from the water source and which are branched into a single path. One of the key features of this figure is that the reliability of the system appears to be almost as high as in the case of completely looped segments. This means that looped segments can secure various paths by which water is supplied, so that rapid service degradation, even in the event of the breakage of some pipes, can be prevented.

To confirm the importance of the water pipes more accurately, 20 major water pipelines were marked as shown in Figure 10. The figure shows that the serviceability of the water pipes in the terminal area, which ultimately branches into a single path, was very low. Therefore, these areas require rehabilitation work to improve the durability or diversification of the water supply paths.

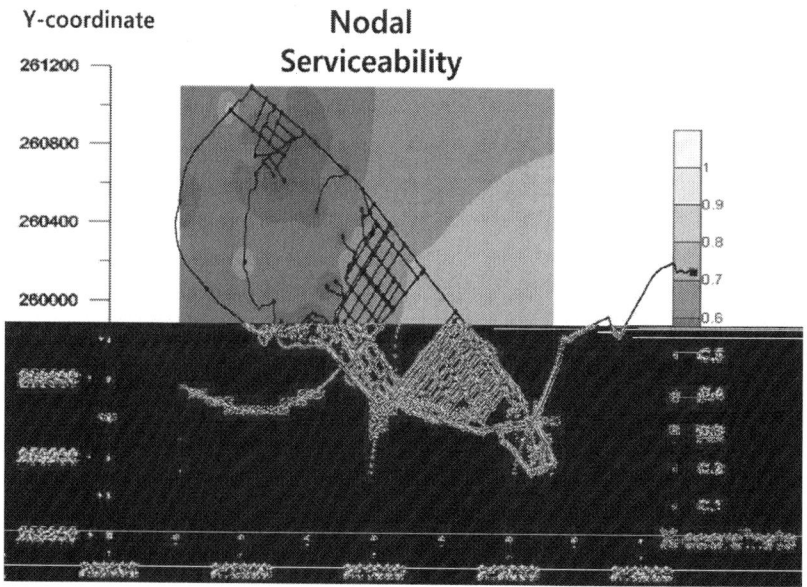

Figure 9. Nodal serviceability of the KA network.

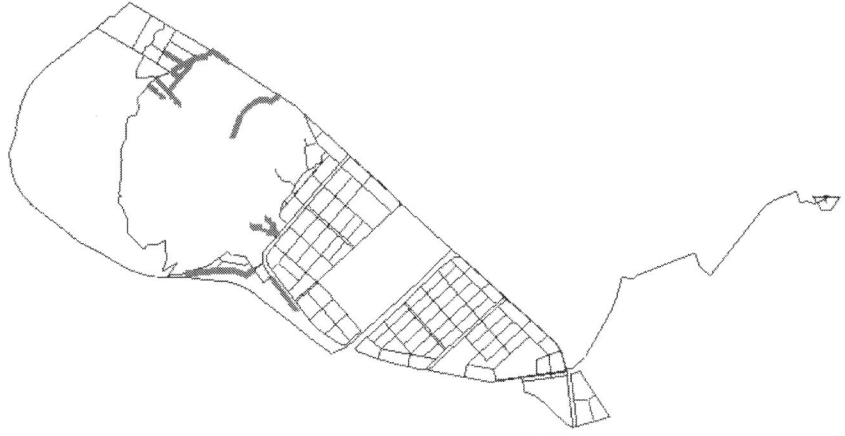

Figure 10. Importance of pipes by multi-pipe failure.

Results for Final Rehabilitation Priority Order

Table 7 summarizes the statistical values of four attributes. Based on the deterioration rate, the relative importance, the weighted utopian approach was used to calculate the rehabilitation priority order of individual water pipes.

Table 7. Statistical value of attributes of deterioration and importance.

Statistical Indicator	Deterioration Rate		Importance	
	Internal Factor	Installation Environments/External Factor	Single Pipe Failure	Multiple Pipe Failure
Average	0.7036	0.1470	0.0832	0.3179
Variance	0.1963	0.0556	0.0090	0.0232
Max value	1.0000	0.7111	0.6340	0.8126
Min value	0.0039	0.0039	0.0001	0.0744

Five weighting methods were used to apply the weighted utopian approach. The weighting factors of the attributes used in four of the methods (excluding the entropic method) were based on the results of the survey of 15 experts in the field of water distribution systems (the results are listed in Table 8). As with the survey concerning the weighting factors for the deterioration rate, the 15 experts consisted of five from academia, five from corporations and five from research centers and government agencies.

Table 8. Weighting factors of attributes depending on weighting methods.

Weighting Method	Weighting Factor			
	Deterioration by Internal Factors	Deterioration by External Factors	Relative Importance by Single Pipe Failure	Relative Importance by Multi Pipe Failure
Same Weighting	0.250	0.250	0.250	0.250
Eigenvector	0.344	0.097	0.349	0.210
Churchman-Ackoff	0.400	0.200	0.300	0.100
Rating	0.351	0.220	0.268	0.162
Entropic	0.155	0.549	0.246	0.050

Table 9 lists the weighting methods used and the resulting priority orders obtained for the rehabilitation of the top 20 water pipes, after applying the WUA. The results obtained using the five weighting methods indicate that the eigenvector, Churchman-Ackoff and rating methods produced similar results, whereas the same weighting and entropic methods produced results that were somewhat different from those obtained with the other three methods. Accordingly, the redundancy of the top 20 water pipes was investigated, and the results confirmed that a total of five water pipes were identified as priorities by all five methods and that a total of 14 water pipes overlapped in more than three of the methods. As described above, the rehabilitation priorities changed slightly as the weights were adjusted. These results indicate that although the superiority of any one of the weighting methods (whereby each of their axiomatic systems has been verified both logically and independently) cannot be determined, the sensitivity to any change in the weighting factors should be taken into account when a decision or rehabilitation prioritization is made. Hence, the proposed model offers not only the advantage of being representative, but also generality among multi-criteria decision-making methods, as it considers several methods of determining weighting factors with diverse characteristics. The proposed method can also provide final total results, as well as the results for each weighting method for consideration by final decision makers, thereby providing a wide range of alternatives depending on the context. That is, if there are attributes that must be considered depending on the decision-making environment and the characteristics of the issues, a specific weighting method can also be used to produce a priority ranking other than the priority ranking produced by the model developed in this study.

In this study, the final rehabilitation priority order was determined based on the average ranking for each water pipe, selected using the five weighting methods. The final results for the top 20 water pipes are shown in Figure 11 and Table 10. Although the results do not exhibit a clear trend, the rehabilitation priority order of the water pipes that are directly connected to the water source or that connect networks tended to receive high priority rankings. Water pipes other than those described above also generally ranked highly in terms of rehabilitation priority because of the effect of the deterioration rate determined from internal factors, for which the weighting values were set high as a result of the survey results. The deterioration rate values resulting from the internal and external factors showed that most of the values for the top 20 water pipes were high and that their variation was not considerable. However, the importance of single and multiple failures showed that the variation in the deterioration rates for the top 20 water pipes was relatively large. This result reflects the characteristics of the distribution of low factor values in most cases, except for some segments, in the case of a single failure. However, in the case of multiple failures, because the weighting values used to calculate the final rehabilitation priority order were not sufficiently large, the effect on the final rehabilitation priority order was minimal, even if the importance value resulting from multiple failures is large, which is why the variation for the top 20 values was large and unevenly distributed.

Table 9. Priority order for rehabilitation according to weighting methods.

Rehabilitation Priority Order	Same Weighting	Eigenvector	Churchman-Ackoff	Rating	Entropic
1	1105	1105	1105	1105	1088
2	1197	975	977	977	1073
3	1236	1104	1104	1197	1236
4	1220	976	975	1220	1105
5	1018	1118	976	1104	1040
6	1153	1005	1118	1018	977
7	909	977	1005	975	986
8	977	1197	1197	1236	1153
9	1121	1220	1220	976	909
10	1241	1009	1018	1088	1132
11	1213	1013	1088	1073	1128
12	1001	1015	1073	1153	1181
13	1132	1014	1206	1118	1197
14	999	1018	937	1206	1206
15	1000	1011	959	909	937
16	1002	1017	950	959	959
17	1226	1019	931	937	950
18	997	1236	930	1005	931
19	995	1206	1097	950	930
20	1199	959	912	930	1097

Figure 11. Top 20 pipes to be rehabilitated.

Table 10. Values of four attributes according to rehabilitation order.

Rehabilitation Priority Order	Deterioration by Internal Factors	Deterioration by External Factors	Relative Importance by Single Pipe Failure	Relative Importance by Multi Pipe Failure
1105	0.9997	0.5187	0.6340	0.2524
977	0.9905	0.5158	0.5538	0.0744
1197	0.9997	0.5332	0.1572	0.6167
1220	0.9975	0.5158	0.1572	0.5729
1236	0.9975	0.7111	0.0624	0.6172
1018	0.9995	0.5158	0.1572	0.4845
1088	0.9978	0.7111	0.1292	0.1686
1073	0.9978	0.7111	0.1292	0.1668
1206	0.9642	0.5332	0.1292	0.2620
959	0.9642	0.5332	0.1292	0.2476
937	0.9997	0.5332	0.1292	0.2426
950	0.9940	0.5332	0.1292	0.2122
930	0.9642	0.5332	0.1292	0.1995
931	0.9905	0.5332	0.1292	0.1960
1153	0.9996	0.7111	0.0039	0.6181
909	0.9998	0.7111	0.0030	0.5858
1104	0.9990	0.0569	0.6340	0.2232
975	0.9998	0.0069	0.6340	0.2908
1097	0.9940	0.5332	0.1292	0.1676
912	0.9940	0.5332	0.1292	0.1670

CONCLUSIONS

To determine the rehabilitation priority order for pipes in a water distribution system, we developed a new method to address the limitations of existing methods that use only the pipeline deterioration rate. We aimed to determine the priority of pipes for rehabilitation consider normal and abnormal conditions using multi-criteria decision making methods (using compromise of some weighting methods). In this study, we also added single and multiple pipe failures simulation in addition to the deterioration of pipes. These are key points of this manuscript. The proposed method considers the hydraulic importance in addition to the deterioration rate in determining the rehabilitation priority order with higher reliability. The results of this study are significant in that they show how a rehabilitation priority order model can be combined with the concept of the hydraulic importance of water pipes in applying existing deterioration rate calculation methods to determine the rehabilitation priority order. The proposed method was applied to the KA water distribution network of City J to determine the rehabilitation priority order for the pipes in the KA network. The results confirm that the proposed method provides a more realistic determination of the rehabilitation priority order that considers not only the deterioration rate, but also the relative hydraulic importance of each pipe. This model is able to determine the rehabilitation priority order for pipes in a water distribution network and can be applied more easily than existing rehabilitation priority order models that require large amounts of data and involve complex failure probabilities and mathematical models.

Because large-scale infrastructure networks, such as water distribution systems, require constant maintenance and rehabilitation, the design and reinforcement of water distribution networks to guard against multiple failures in the event of events, such as earthquakes, place huge financial and physical burdens on the water supplier. Korea has never experienced large-scale earthquakes, and most Koreans believe that their country has little to fear in this respect. However, although the probability of a disaster, such as an earthquake in Korea, is low, such an event would cause significant damage if it were to occur. Thus, Korea needs to be prepared for such disasters. Given these circumstances, a method that can determine the level of reliability of a water distribution system in the face of a disaster capable of causing immense damage is necessary within the current framework of maintenance and rehabilitation. The proposed model is also advantageous from this perspective.

The results of this study suggest that further research on this subject is warranted. The proposed method should be compared with other recent methodologies, such as CARE-W, CASSES and AWARE-P. Therefore, a detailed comparative study can be one of the further studies in the near future. The possible malfunction of valves was not considered in calculating the relative importance of structural failures. Because the area in which damage may occur varies depending on the ability to operate the relevant valve, the effect of valve malfunctioning should be studied. In this study, given that water distribution systems are essential social infrastructure networks that directly affect public welfare; an economic feasibility analysis for cost optimization purposes was not performed. Therefore, further studies on rehabilitation prioritization with consideration of economic constraints should be undertaken.

ACKNOWLEDGMENTS

This work was supported by a grant from the National Research Foundation (NRF) of Korea, funded by the Korean government (Ministry of Science, ICT and Future Planning, MSIP) (No. 2013R1A2A1A01013886).

AUTHOR CONTRIBUTIONS

Do Guen Yoo and Doosun Kang carried out the survey of previous studies, analysis of proposed method, participated in the sequence alignment and drafted the manuscript; Joong Hoon Kim and Hwandon Jun conceived the original idea of the study, and helped to write the final manuscript.

REFERENCES

1. American Water Works Association Research Foundation (AWWARF). *Guidance Manual Water Mains Evaluation for Rehabilitation/Replacement*; AWWARF: Denver, CO, USA, 1986.
2. Water Research Centre (WRC). *Planning the Rehabilitation of Water Distribution Systems*; WRC: Marlow, UK, 1989.
3. K-Water. *Development of Decision-Making System for Pipe Rehabilitation*; K-Water: Daejeon, Korea, 1995.

4. Kim, E.S.; Baek, C.W.; Kim, J.H. Estimate of Pipe Deterioration and Optimal Scheduling of Rehabilitation. *Water Sci. Technol. Water Supply* **2005**, *5*, 39–46.
5. Shamir, U.; Howard, C.D. An Analytic Approach to Scheduling Pipe Replacement. *J. Am. Water Works Assoc.* **1979**, *71*, 248–258.
6. Marks, D.H.; Clark, M.R. A New Methodology for Modeling Break Failure Patterns in Deteriorating Water Distribution Systems: Theory. *Adv. Water Resour.* **1987**, *10*, 2–10.
7. Marks, D.H.; Clark, M.R. A New Methodology for Modelling Break Failure Patterns in Deteriorating Water Distribution Systems: Applications. *Adv. Water Resour.* **1987**, *10*, 11–20.
8. Agbenowosi, N.K. A Mechanistic Analysis Based Decision Support System for Scheduling Optimal Pipeline Replacement. Ph.D. Thesis, Virginia Polytechnic Institute and State University, Blacksburg, VA, USA, September 2000.
9. Park, S.; Loganathan, G.V. Methodology for Economically Optimal Replacement of Pipes in Water Distribution Systems: 1. Theory. *KSCE J. Civil Eng.* **2002**, *6*, 539–543.
10. Park, S.; Loganathan, G.V. Methodology for Economically Optimal Replacement of Pipes in Water Distribution Systems: 2. Application. *KSCE J. Civil Eng.* **2002**, *6*, 545–550.
11. Deb, A.K.; Hasit, Y.J.; Grablutz, F.M.; Herz, R.K. *Quantifying Future Rehabilitation and Replacement Needs of Water Mains*; Report to the American Water Works Association Research Foundation. Roy F. Weston, Inc.: West Chester, PA, USA, 1997.
12. Walski, T.M. Economic Analysis of Rehabilitation of Water Mains. *J. Water Resour. Plan. Manag.* **1982**, *108*, 296–304.
13. Luong, H.T.; Fujiwara, O. Fund Allocation Model for Pipe Repair Maintenance in Water Distribution Networks. *Eur. J. Oper. Res.* **2002**, *136*, 403–421.
14. Alvisi, S.; Franchini, M. Near Optimal Rehabilitation Scheduling of Water Distribution Systems based on Multi-objective Genetic Algorithms. *Civil Eng. Environ. Syst.* **2006**, *23*, 143–160.
15. Saegrov, S. *Computer Aided Rehabilitation of Water Networks*; IWA Publishing: London, UK, 2005; ISBN 1843390914.
16. National Civil Engineering Laboratory. *CARE-W WP1 D1 Construction of Control Panel for Performance Indicators for Rehabilitation*; EVK1-CT-2000–00053, No. 1.1. Research and Technological Development Project of European Community, European Commission: Brussels, Belgium, 2001.

17. National Civil Engineering Laboratory. *CARE-W WP1 D2 Description of Technical Tools for Failure Forecasting and Network Reliability*; EVK1-CT-2000–00053, No. 1.2. Research and Technological Development Project of European Community, European Commission: Brussels, Belgium, 2002.
18. Cemagref Bordeaux. *CARE-W WP2 D3 Description and Validation of Technical Tools: Models Description*; EVK1-CT-2000–00053, No. 2.1. Research and Technological Development Project of European Community, European Commission: Brussels, Belgium, 2002.
19. INSA Lyon-URGC. *CARE-W WP3 D6 Decision Support for Annual Rehabilitation Programs: Criteria for the Prioritization of Rehabilitation Projects*; EVK1-CT-2000–00053, No. 3.1. Research and Technological Development Project of European Community, European Commission: Brussels, Belgium, 2002.
20. INSA Lyon-URGC. *CARE-W WP3 D6 Decision Support for Annual Rehabilitation Programs: Survey of Multi-Criteria Techniques and Selection of Relevant Procedures*; EVK1-CT-2000–00053, No. 3.2. Research and Technological Development Project of European Community, European Commission: Brussels, Belgium, 2002.
21. Technique University Dresden. *CARE-W WP4 D9 Development of the "Rehab Strategy Manager" Software*; EVK1-CT-2000–00053, No. 4.2. Research and Technological Development Project of European Community, European Commission: Brussels, Belgium, 2002.
22. Cemagref (2008) CASSES, User Manual. Available online: http://casses.irstea.fr/wp-content/uploads/2013/02/CassesManual_2.0.0_b.pdf (accessed on 1 November 2014).
23. AWARE-P (2012). The AWARE-P project—Infrastructure asset management of urban water services. Impact report for application to the 2012 Muelheim Water Award. Available online: http://www.aware-p.org/np4/?newsId=13&fileName=MWA_AWARE_P_Application_compressed.pdf (accessed on 1 November 2014).
24. Saaty, T.L. *The Analytic Hierarchy Process*; McGraw-Hill: New York, NY, USA, 1980.
25. Specht, D.F. Probabilistic Neural Network. *Neural Netw.* **1990**, *3*, 109–118.
26. Jun, H.; Loganathan, G.V.; Kim, J.H.; Park, S. Identifying Pipes and Valves of High Importance for Efficient Operation and Maintenance of Water Distribution Systems. *Water Resour. Manag.* **2008**, *22*, 719–736.
27. Walski, T.M. Water Distribution Valve Topology for Reliability Analysis. *Reliab. Eng. Syst. Saf.* **1993**, *42*, 21–27.

28. Yoo, D.G.; Kang, D.; Kim, J.H. Seismic Reliability Assessment Model of Water Supply Networks. In Proceedings of the World Environmental and Water Resources Congress, Cincinnati, OH, USA, 19–23 May 2013; pp. 955–966.
29. Yoo, D.G.; Kang, D.; Kim, J.H. Applications of Seismic Reliability Assessment Model for Water Distribution Networks. In Proceedings of the World Environmental and Water Resources Congress, Portland, OR, USA, 1–5 June 2014.
30. Cullinane, M.J.; Lansey, K.E.; Mays, L.W. Optimization-Availability-Based Design of Water-Distribution Networks. *J. Hydraul. Eng.* **1992**, *118*, 420–441.
31. Tabucchi, T.; Davidson, R.; Brink, S. Simulation of Post-Earthquake Water Supply System Restoration. *Civil Eng. Environ. Syst.* **2010**, *27*, 263–279.
32. Lansey, K.E. Sustainable, Robust, Resilient, Water Distribution Systems. In Proceedings of the 14th Water Distribution Systems Analysis Symposium, Engineers Australia, Adelaide, Australia, 24–27 September 2012.
33. Knoll, A.L.; Engelberg, A. Weighting Multiple Objectives—The Churchman-Ackoff Techniques Revisited. *Comput. Oper. Res.* **1978**, *5*, 165–177.
34. Eckenrode, R.T. Weighting Multiple Criteria. *Manag. Sci.* **1965**, *12*, 180–192.
35. Hwang, C.L.; Yoon, K. *Multiple Attribute Decision Making Methods and Applications: A State-of-The-Art Survey*; Springer-Verlag: New York, NY, USA, 1981.
36. Soofi, E.S. Generalized Entropy-Based Weight for Multi-Attribute Models. *Oper. Res.* **1990**, *32*, 362–363.
37. Xanthopulos, Z.; Melachrinoudis, E.; Solomon, M.M. Interactive Multiobjective Group Decision Making with Interval Parameters. *Manag. Sci.* **2000**, *46*, 1721–1732.
38. Yoo, D.G.; Kim, J.H.; Jun, H. *Determination of Rehabilitation Priority Order of Subareas in Water Distribution Systems Considering the Relative Importance of Pipes*; Water Distribution Systems Analysis 2010. Americal Society of Civil Engineers: Reston, VA, USA, 2010; pp. 1045–1052.
39. Korea Water and Wastewater works Association (KWWA). *Water Supply Facility Standard*; Ministry of Environment: Seoul, Korea, 2010.

CITATION

Do Guen Yoo, Doosun Kang, Hwandon Jun and Joong Hoon Kim, Rehabilitation Priority Determination of Water Pipes Based on Hydraulic Importance, doi:10.3390/w6123864.

CHAPTER 8

Evaluation of Actions for Better Water Supply and Demand Management in Fayoum, Egypt Using Ribasim

Mohie M. Omar

National Water Research Center, Ministry of Water Resources and Irrigation, Shoubra El-Kheima, Egypt

ABSTRACT

Fayoum Governorate faces many water-related challenges being; compensating the water shortage and controlling the volumes of drainage water effluents into Quarun Lake. There are many actions, based on water resources management approach, which can help overcome these water-related challenges. These actions are classified to developing additional water resources (supply management), and properly using the existing water resources (demand management). This study investigates using the RIBASIM (RIver BAsin SIMulation) model, the most suitable actions for the future. RIBASIM was used to simulate the current condition and evaluate various scenarios in 2017 based on different actions. Three scenarios were formulated being optimistic, moderate, and pessimistic which represent different implementation rates of the tested actions. RIBASIM results indicated a water shortage of 0.59, 1, and 1.85 Billion Cubic Meter (BCM)/year, for the simulated scenarios, respectively. Since Fayoum is a miniature of Egypt with respect to both, the natural and water resources systems, the results of this study can be used as guidelines for optimization of the water resources system in Egypt.

INTRODUCTION

Fayoum Governorate as shown in Fig. 1 and Fig. 2, is a large depression in the Egyptian western desert, located 90 km south-west of Cairo. Bahr Youssef Canal is the only water resource for Fayoum, and it is one of the main branches of Ibrahimia Canal. Ibrahimia Canal takes off the water from the Nile River at a distance of 539 km from Aswan High Dam. The length of Bahr Youssef Canal is 313 km and it provides water to four districts: West Menia, Bani-Swif, Fayoum and Giza. Fayoum is the largest district served by Bahr Youssef Canal with 454,700 Feddans (1 Feddan = 4200 m^2) (NWRP/MWRI, 2013). Fayoum Governorate has been selected in this study, because it has similar characteristics of Egypt with respect to both, the natural and water resources systems. Fayoum water shortage is compensated by drainage reuse, which negatively affects the soil and plants. The remaining drainage water flows into Quarun Lake and Rayan Channel. The excess of drainage water beyond the capacity of Quarun Lake and Rayan Channel floods the surrounding villages, lands, and resorts. This limits horizontal expansion projects, since Fayoum is a closed depression.

Figure 1. Location of Fayoum in Egypt (National Water Research, Egypt).

INTRODUCTION

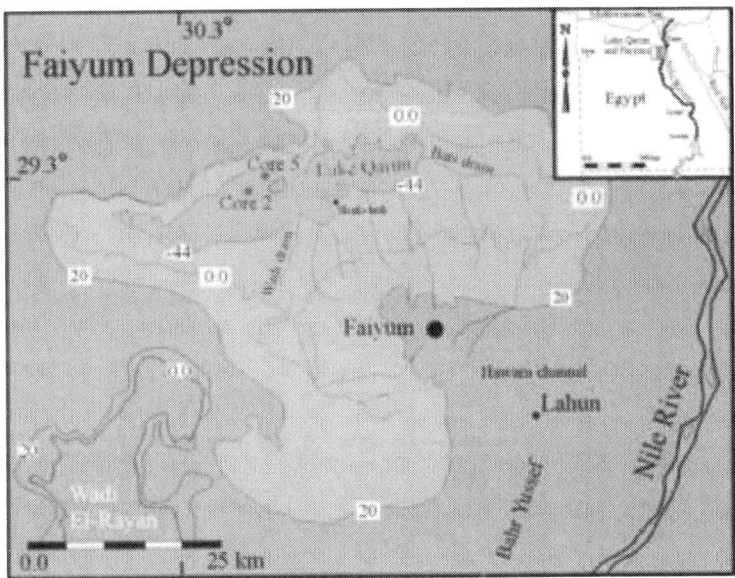

Figure 2. Fayoum depression, Lake Quarun and the surrounding deserts in 2012.

This primary goal of this study is to evaluate the influences of different management actions on the quantitative water system performance of Fayoum Governorate in the future. This evaluation enables the interested stakeholders to identify and implement actions for minimizing water shortages and controlling volumes of drainage water effluents into Quarun Lake.

It was found that many researchers locally and worldwide investigated the water resources management actions using numerical models. The Agro-hydrological modeling system (ACRU) has been developed and applied in South Africa for simulation of land use/management influences on water resources demand and supply (Schulze and Smithers, 2004). Water evolution and planning (WEAP) model has been developed by Stockholm environment Institute to evaluate planning issues related to water resources for both municipal and agricultural systems including: sector demand analyses, water conservation, water rights and allocation priorities, stream flow simulation, reservoir operation, ecosystem requirements and project cost–benefit analyses. The model has been applied to assess scenarios of water resource development in the Pangani Catchment in Tanzania (Arranz and McCartney, 2007).

The Nile Decision support Tool (Nile DST) has been developed as part of the FAO Nile Basin Water Resources project to objectively assess the benefits and tradeoffs associated with various water development strategies. The Nile DST comprises six main components: databases, river simulation and management agricultural planning, hydrologic modeling, remote sensing and user-model interface (Georgakakos, 2006).

RIBASIM was applied in more than 20 countries to support the process of water resources planning. Van der Krogt and Verhaeghe (2001) used the RIBASIM Model to describe the effects of changes in the farming system on the regional water balance for three river basins via the Jratunseluna, Serayu and Cidurian Basins, Indonesia. Van der Krogt (2010) applied the RIBASIM model to clarify the importance of the natural system for the socioeconomic situation, and also to develop the Sistan Basin, Iran. Through the Hydrology Project at Water Resources Department Government of Maharashtra, India, RIBASIM model was used to predict the water shortages in the Godavari river basin, India for years 2015 and 2020, and to develop decisions for minimizing deficit.

The National Water Resources Plan in Egypt (NWRP) developed a Decision Support System (DSS) based on the RIBASIM7 model. The author was involved in developing the NWRP-DSS. RIBASIM7 of NWRP-DSS provided a full picture of the current water balance in Egypt at Aswan High Dam. But, it did not show the future situations according to expected developments and different management measures.

METHODOLOGY

The study area
Several visits were carried out to Fayoum in order to perceive a complete data picture of the study area. The visits included two site locations being; Quarun lake, and one of the new agricultural projects in the surrounding desert eastern of Fayoum depression. The visits also included the Water Resources Unit, Fayoum irrigation directorate, Fayoum agricultural directorate, the holding company for water and wastewater treatment at Fayoum. The collected data were the population number and population growth, total agricultural area and cropping patterns, total agricultural area of new lands in the surrounding deserts and the applied irrigation systems, capacities of all drinking water plants, capacities of primary and secondary wastewater treatment plants, number of factories and total industrial demand, and the total irrigation volumes discharged into agricultural lands.

METHODOLOGY

Based on the site visits and the assembled data, Fayoum water balance was developed for water resources analysis (Fig. 3). The water balance is an accounting of the inputs and outputs. It also represents the water consumptions for all sectors, and the volumes which return back to the system. Fayoum water balance could be described, as follows:

- The major inputs are the discharge from Bahr Youssef Canal via Lahon Dam, rainfall, and shallow groundwater;
- The outputs are evaporation, water uses for different sectors and drainage to Quarun Lake and Rayan Valley;
- The total domestic and industrial demand is 0.287 BCM, which is obtained as the total capacity of all drinking water plants and factories. The actual use is only 0.07 BCM, which is calculated as population number multiplied by consumption rate. The remaining volume returns back to the system as treated wastewater (0.117 BCM) and untreated wastewater (0.1 BCM).

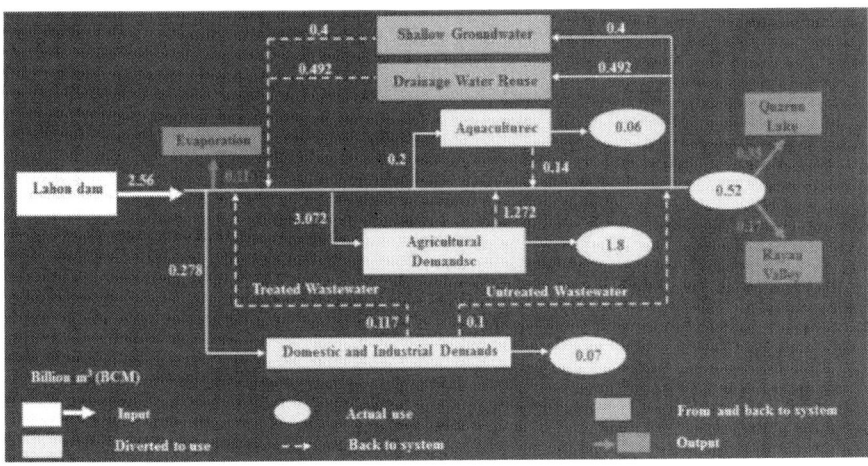

Figure 3. Water balance in Fayoum in 2011.

The agricultural demand is 3.072 BCM, which is obtained as the total irrigation volumes discharged from canals to agricultural lands via weirs. The actual consumption is 1.8 BCM, which is calculated as the sum of each crop multiplied by its consumption rate. The remaining volume of 1.272 BCM returns back to the system.

The geomorphologic features of Fayoum depression are Ecocene limestone and marls, surrounded on all sides by high rocky walls. These formations generally contain no groundwater aquifers. There is also the

Nubian sandstone aquifer whose depth is too great to allow exploitation. Thus, groundwater is unreliable source in Fayoum. The annual rainfall averages 14 mm, and is rather irregular in time and place. Thus, rainfall is very scanty and cannot be a reliable source of water. The soil analyses of Fayoum Governorate show that the soil texture is clay in Fayoum depression, and sandy loam in the desert surrounding the depression (Shendi et al., 2006).

- Based on the site visits, assembled data and water balance analyses, the challenges facing the study area could be described, as follows:
- The water availability is limited, since Fayoum is a closed water depression with only one limited source of water from Bahr Youssef Canal with very scanty rainfall, and no reliable groundwater.
- The drainage water is the limiting factor in the water balance which controls the irrigation water requirements. This is attributed to the fact that Quarun Lake is of limited capacity that receives about 67% of the total drainage water (i.e. 0.35 BCM/year). Above this amount, the lake floods over the surrounding villages, lands, and resorts which limits the expansion of irrigated agriculture.
- The soils of the new agricultural lands in the desert areas surrounding Fayoum depression have a low water storage capacity and a high infiltration rate due to its sandy loam texture. In addition, old irrigation methods are applied in these lands. This means that the field application loss is high in these new agricultural lands.

From the above, the main problem facing Fayoum is the excess of drainage water beyond the capacity of Quarun Lake. This deteriorates the soil and plants' productivity. If the drainage water into Quarun Lake exceeds 0.35 BCM/year, the water level rises in the lake and floods the surrounding villages, lands, and resorts. The last accident recorded was in November 2012, when the lake flooded over the agricultural lands and houses of two villages being Shakshook and Shamata located on the western coast of the lake.

RIBASIM description and model capability

It was decided to rely on RIBASIM model in order to evaluate different management actions that can contribute in minimizing the water shortage and reducing the drainage water effluent to Quarun Lake. RIBASIM is a generic model package for simulating the behavior of river basins under various hydrological conditions. The model package is a comprehensive and flexible tool which links the hydrological water inputs at various locations with the specific water-users in the basin.

The structure of RIBASIM model is based on an integrated framework with a user-friendly, graphically oriented interface. The main RIBASIM user interface is a flow diagram representing the tasks involved in carrying out a simulation analysis is used to assist the user through the analysis from data entry to the evaluation of the results. As for the capabilities of RIBASIM, it can model various future and potential situations and system configurations by setting various scenarios and management actions (Van der Krogt, 2010).

Schematization of RIBASIM

The schematization of RIBASIM reproduces all necessary features of a basin by nodes connected by links. Van der Krogt (2010) defined the model schematization as a translation and a simplification of the real world into a format which allows the actual simulation. The result of the schematization is a network of nodes and links which reflects the spatial relationships between the elements of the basin, and the data characterizing those nodes and links. The nodes represent reservoirs, dams, weirs, pumps, hydro-power stations, water users, inflows, man-made and natural bifurcations, intake structures, and natural lakes. The links transport water between the different nodes.

Schematization of water system of Fayoum Governorate

For Fayoum Governorate, groundwater was not modeled in the schematization. This is because deep ground water is rarely available and irrigates only 0.15% of the total agricultural area due to the geologic nature of the place. Shallow groundwater is directly used by the local water users for irrigation, so it is considered as a reuse of drainage water. All the necessary features of Fayoum water system were reproduced by 11 nodes and 13 links.

The "Advanced IRrigation (AIR) node" reflected the water demand for irrigation. Only one "Advanced irrigation node" was located for Fayoum. The water demand was computed based on crop characteristics and crop plans. The crop characteristics were stored in the fixed data base of RIBASIM. However, the crop plan was generated by the Agricultural Sector Model for Egypt (ASME). The ASME model can be used to estimate the annual crop plan for a specific area (MWRI, 2001). Therefore, the ASME model was also used in this study to compute the annual crop plan (list of cultivations), which was developed as direct inputs to RIBASIM. Both RIBASIM and ASME used the same 31 crops. The crop characteristics in the fixed data base of RIBASIM7 were described as crop plan and crop characteristics, hydrological input, soil characteristics, topography and lay-out of the irrigation area, operation and irrigation management, and potential crop yield and production costs, and actual field water balance.

The "Public Water Supply (PWS) node" represented the domestic, municipal and industrial demands. For domestic demand, the population and consumption rates were input values to RIBASIM7. The domestic demand in Fayoum was divided into two PWS nodes. One node represented the demand from Bahr Youssef Canal and its branches while the second node represented the demand from Bahr Hassan Wasef Canal and its branches. Bahr Hassan-Wasef is a main canal taking its water from Bahr Youssef Canal. Table 2 presents the population number and the consumption rate for both PWS nodes. The industrial demand was also represented by one PWS node with an explicit demand of 0.63 m^3/s.

Table 1. Water balance in Fayoum in the year 2011.

Water supply				Water demand		
Conventional resources		Unconventional resources		Sector	Estimated consumption (BCM/year)	Actual use with losses (BCM/year)
Water Resources	Volume (BCM/year)	Water Resources	Volume (BCM/year)	Municipal and industry	0.07	0.287
Lahon Dam	2.56	Shallow Groundwater	0.4	Irrigation	1.8	3.072
		Drainage Water Reuse	0.492	Aquaculture	0.06	0.2
		Wastewater	0.217	Evaporation	0.11	0.11
Total	2.56	Total	1.109	Drainage to Quarun lake	0.35	
				Drainage to Rayan Valley	0.17	
Total	3.669	Total	3.669	Total	2.56	

METHODOLOGY

Table 2. PWS nodes for domestic demand in Fayoum.

Node name	Description	Population (–)	Unit demand (l/capita/day)
Dom_FAY_FAY_Fayoum1	From Bahr Youssef canal and its branches serving the middle and the northern east parts of Fayoum	2,086,350	175
Dom_FAY_FAY_Fayoum2	From Bahr HassanWasef canal and its branches serving the southern and western parts	460,532	175

There was one main diversion node which reflected the Lahon Dam, where the water entered Fayoum Governorate from Bahr Youssef Canal. There were other three diversion nodes representing the water diverting into one irrigation node (AIR) and two domestic (PWS) nodes.

There was only one recording node to represent the flow gauging station for Fayoum network. The recording node followed Lahon Dam diversion node was used for comparison between simulated demand and monitored flow (supply) at Lahon Dam. One terminal node was available in the schematization representing the downstream boundary of the system where water leaves the network. One confluence node was allocated in the schematization to represent the location where various outflows of different nodes join the system.

There were four diversion links in the schematization being; one diverted flow from Lahon Dam, one diverted irrigation flow, and two diverted flows for domestic demands. In addition, there were eight surface water flow links used to link between any different nodes, except those links following diversion nodes.

Simulated scenarios

The strategy of Fayoum Water Resources Plan-2017 is a coherent combination of actions with respect to water quantity and quality. The current study only focused on actions dealing with water quantity including developing new water resources (supply management), and properly using existing water resources (demand management). It is to be noted that the sensitivity analysis of actions in future scenarios requires modifications of some input data in RIBASIM. Table 3 shows the input values for the current scenario and the future scenarios being optimistic, moderate, and pessimistic. The descriptions of all scenarios are given in the following sections.

Table 3. Tested actions and their corresponding inputs in RIBASIM7.

Tested actions	Modified inputs in RIBASIM7	Current scenario (actual)	Optimistic scenario (assumed)	Moderate scenario (assumed)	Pessimistic scenario (assumed)
Increase fresh water availability from Nile at Lahon Dam	Time series monitored flow (10-day step) in the TMS file a	10-day flow (m3/s) in the TMS file of current situation with a sum of 2.56 BCM/year	3.119 BCM/year is distributed over 10-day values (m3/s) with same patterns of the current TMS file	2.777 BCM/year is distributed over 10-day values (m3/s) with same patterns of the current TMS file	2.579 BCM/year is distributed over 10-day values (m3/s) with same patterns of the current TMS file
Continue improvement irrigation project	Distribution efficiency in advanced irrigation node	56%	70%	63%	56%
Maintenance of canals with high losses	Conveyance efficiency in advanced irrigation node	56%	70%	63%	56%
Apply modern irrigation techniques	Field application efficiency in advanced irrigation node	56%	70%	63%	56%
Make horizontal agricultural expansions	Total area in the advanced irrigation node	512,000 Feddan (215,040 ha)	570,000 Feddan (239,400 ha)	560,000 Feddan (235,200 ha)	545,000 Feddan (228,900 ha)
Make campaigns to reduce average of rural and urban consumption	Demand in the public water supply node	195 l/capita/day	185 l/capita/day	190 l/capita/day	195 l/capita/day
Promote domestic water saving technologies	Distribution loss in public water supply node	30%	20%	25%	30%

METHODOLOGY 195

Tested actions	Modified inputs in RIBASIM 7	Current scenario (actual)	Optimistic scenario (assumed)	Moderate scenario (assumed)	Pessimistic scenario (assumed)
Increase fresh water availability from Nile at Lahon Dam	Time series monitored flow (10-day step) in the TMS file[a]	10-day flow (m³/s) in the TMS file of current situation with a sum of 2.56 BCM/year	3.119 BCM/year is distributed over 10-day values (m³/s) with same patterns of the current TMS file	2.777 BCM/year is distributed over 10-day values (m³/s) with same patterns of the current TMS file	2.579 BCM/year is distributed over 10-day values (m³/s) with same patterns of the current TMS file
Continue improvement irrigation project	Distribution efficiency in advanced irrigation node	56%	70%	63%	56%
Maintenance of canals with high losses	Conveyance efficiency in advanced irrigation node	56%	70%	63%	56%
Apply modern irrigation techniques	Field application efficiency in advanced irrigation node	56%	70%	63%	56%
Make horizontal agricultural expansions	Total area in the advanced irrigation node	512,000 Feddan (215,040 ha)	570,000 Feddan (239,400 ha)	560,000 Feddan (235,200 ha)	545,000 Feddan (228,900 ha)

Make campaigns to reduce average of rural and urban consumption	Demand in the public water supply node	195 l/capita/day	185 l/capita/day	190 l/capita/day	195 l/capita/day
Promote domestic water saving technologies	Distribution loss in public water supply node	30%	20%	25%	30%

^aTMS: time series file producing monitored (supply) flow at the recording node of Lahon dam.

Current scenario

Nile River is the only source of water from Bahr Youssef canal with 2.56 BCM/year. Due to the geomorphologic and climatic features of Fayoum depression as explained in Section 2.1, the groundwater and rainfall are is unreliable sources.

Concerning agricultural demands, agriculture is the largest consumer of water which irrigates 407,544 Feddan (169,810 ha) in Fayoum. The irrigation efficiency for the whole agricultural network is 56%, calculated as the ratio of the amount reaching the root zone of the plants (estimated consumption) to the amount diverted to the system (actual use with losses). The irrigation efficiency is subdivided into field application, distribution, and conveyance efficiencies. Therefore, the values of field application, distribution, and conveyance efficiencies for the current scenario in RIBASIM were assumed to be 56%.

Concerning domestic and industrial demands, inputs in RIBASIM7 are the total population number (2.781 million capita), population growth rate (2.42%), average consumption use of water for rural and urban population (195 l/capita/day) and distribution water loss (30%).

Future scenarios

Three scenarios were formulated optimistic, moderate, and pessimistic to represent different rates of implementation of the tested actions in the year 2017. The tested actions were classified under two main pillars which are: (i) developing additional water resources (supply management), and (ii) properly using existing water resources (demand management).

Optimistic scenario
Regarding the supply management, three main projects are expected to be fully implemented which will provide additional water quantities from the Nile river to Fayoum Governorate. The three projects are as following:

I. (i) New Bahr Kouta Project which takes off an amount of 0.36 BCM/year from Ibrahimia canal from a distance of 25 km before Lahon dam to feed Bahr Kouta canal;
II. (ii) Bahr Gerza Project which intakes an amount of 0.019 BCM/year directly from the Nile river via pipes that enter Fayoum Governorate at Gerza village and then to Tamia district;
III. (iii) Bahr Wahby Project which supplies an amount of 0.18 BCM/year from Ibrahimia canal to Bahr Wahby canal.

Therefore, the water amount of the Nile River to Fayoum Governorate was 3.12 BCM/year in this scenario.

For agricultural demand management, more efforts are expected toward decreasing the areas of crops having high consumption rates of water such as rice. It is also expected to accelerate the implementation rates of irrigation improvement projects (IIP). The projects include actions for increasing the irrigation network efficiency such as installing automatic downstream water level control gates, and maintenance of branch and distributary Canals. The irrigation methods in most of new agricultural lands in Fayoum desert areas with sandy and loam sandy soils are old methods. This is not in agreement with Brouwer et al. (1988), which recommended that the sandy and loam sandy soils need frequent but modern small irrigation applications such as sprinkler or drip irrigation. Therefore, a wide application of such irrigation techniques is expected in this scenario, which improves the field application efficiency.

As a result, the water use efficiency for the whole agricultural network is expected to increase up to 70%. Therefore, the values of field application, distribution, and conveyance efficiencies in RIBASIM were modified to 70% in the optimistic scenario instead of 56% in the current situation. It is also assumed that growth rate of agricultural lands will increase by 11,000 Feddan/year due to horizontal expansion plans and will decrease by 0.5 Feddan/year due to urbanization. As a result, the total agricultural area is to 575,000 Feddan (239,400 ha) as an input in this scenario.

For domestic and industrial demand management, a successful public awareness campaign could reduce population growth rate to 2.1%. This makes the population number 3.26 million capita in this scenario. The campaign could also reduce the water consumption rate to 185 l/capita/day. It is also expected to achieve a progress in application of water saving technologies in municipal and industrial sectors. This could make a water distribution loss reduction by 20%.

Moderate scenario
For supply side, Bahr Gerza and Bahr Wahby Projects are expected to be completely implemented. However, New Bahr Kouta Project is assumed to provide only half of the targeted quantity due to the conflict with the neighboring governorates around Bahr Youssef Canal (the source of the project). Therefore, the water amount of the Nile River to Fayoum Governorate is 3.12 BCM/year in this scenario.

For agricultural demand management, areas of rice and other crops having high consumption rates of water are expected to be less than them in the optimistic scenario. It is also expected to have moderate implementation rates of irrigation improvement projects (IIP). Application of sprinkler or drip irrigation methods is also planned in the new agricultural lands but in smaller areas than those in the optimistic scenario. As a result, the water use efficiency for the whole agricultural network is expected to increase up to 63%.

Therefore, the values of field application, distribution, and conveyance efficiencies in RIBASIM are modified to 63%. It is assumed that growth rate of agricultural lands will increase only by 9000 Feddan/year due to horizontal expansions and will decrease by 1000 Feddan/year due to urbanization. Hence, the total area is 560,000 Feddan (235,200 ha) in this scenario.

For domestic and industrial demands, a successful public awareness campaign could make the population growth rate to be 2.25% which makes the population number becomes 3.3 million capita. The campaign also reduces the consumption to 190 l/day/capita. The distribution water loss is assumed to decrease to 25% as a result of a little progress in application of water saving technologies.

Pessimistic scenario
Only Bahr Gerza project is expected to be fully implemented. This project will add extra amount of Nile river water, which will raise the total amount to 2.579 BCM/year. It is assumed in this scenario that the other planned projects will stop due to conflicts with neighboring governorates or due to financial obstacles.

For agricultural demands, no success is expected in decreasing rice areas and other crops having high consumption rates of water. Very low implementation rates of irrigation improvement projects (IIP) are expected. Application of sprinkler or drip irrigation methods in the new agricultural lands will remain limited in small areas. As a result, the water use efficiency for the whole agricultural network will be equal to the current efficiency (56%).

Therefore, the values of field application, distribution, and conveyance efficiencies in RIBASIM7 are 56%. It is also assumed that growth rate of agricultural lands to increase by 7000 Feddan/year due to horizontal expansions and decrease by 1500 Feddan/year due to the urbanization. Therefore, the total agricultural area is 545,000 Feddan (228,900 ha).

For domestic and industrial demands, the absence of public awareness campaign makes the population growth rate 2.4% resulting in 3.325 million capita. The consumption use of water remained 195 l/day/capita. The distribution water loss is assumed to remain 30% as a result of expected failure in application of water saving technologies for municipal and industrial sectors.

RESULTS

Data was applied to the above scenarios and results were obtained. The results are analyzed and represented. This section is devoted to present the results. For the verification of the base case 2011, Fig. 4 shows the simulated water demand for Fayoum at the recording node of Lahon dam every 10-day time step along year 2011 by RIBASIM. It also shows the actual measured demand obtained from Fayoum Water Resources Plan, in which the author participated in preparation.

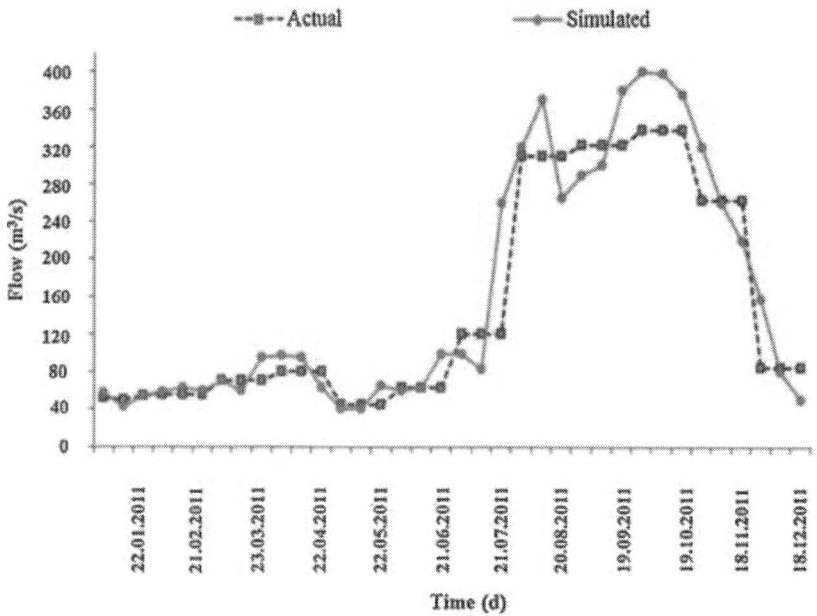

Figure 4. Actual (measured) and simulated (RIBASIM7) demand in Fayoum during year 2011.

The accuracy of the numerical model predictions for the current scenario was measured in terms of the Root-mean-square deviation (RMSD), which was calculated as follows:

$$\text{RMSD} = \sqrt{\frac{\sum_{i=1}^{n}(\hat{Y}_i - Y_i)}{n}} \quad (1)$$

where $\hat{Y}_i$ is a vector of a number (n) of simulated values, Y_i is the vector of a number (n) of actual values.

The RMSD value was 29.5, which was statistically considered low. This value indicated that the RIBASIM model can perform well in the evaluation process of the three future scenarios (optimistic, moderate, and pessimistic).

It was also noticed that the total simulated demand was 3.91 BCM/year which was obtained as the area under the demand curve in Fig. 4. However, the total actual demand was 3.669 BCM/year, as estimated from the collected data from Fayoum visits and the water balance in Table 1.

RESULTS

At the recording node of Lahon dam, the water supply values were entered as inputs in the 10-day time series (TMS) file, and the water demand values were simulated. When the yearly water supply was changed due to planned projects, a new TMS file was developed with new 10-day values distributed as same ratios as patterns of the current file. Fig. 5 shows both, the water supply and simulated demand at Lahon dam distributed over 10-day time steps. Based on the difference between the supply and demand, the current water shortage was 1.6 BCM/year without the reuse of drainage water or wastewater. It was observable, that the water shortage was accumulated in the period between July and November.

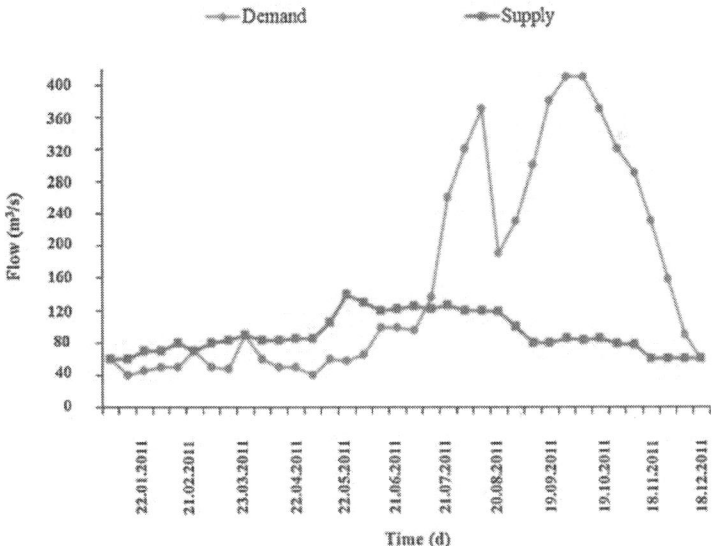

Figure 5. Simulated demands and supply in Fayoum during the year 2011 without the reuse of drainage water or wastewater.

As for future scenarios, the water shortages were also presented without the reuse of drainage water or wastewater (Table 4). The yearly water shortages in the optimistic scenario, moderate scenario, and pessimistic scenario will be 0.59, 1, and 1.85 BCM/year, respectively. The water shortages in the three scenarios will be clearly found in the period between July and November. The water shortages will be compensated by drainage water reuse. It is noticed that, there will be a high excess in water supply from January to June in the three future scenarios. The excessive volumes will be 1.25, 0.9, and 0.35 BCM for the optimistic, moderate, and pessimistic scenarios, respectively. The estimated water balance for future scenarios showed that, the drainage water effluent into Quarun Lake will be 0.35, 0.34, and 0.32 BCM/year in the optimistic, moderate, and pessimistic scenarios, respectively.

Table 4. Water supply and simulated demands (m³/s) every 10-day time steps for the three scenarios during the year 2017.

Pessimistic scenario	Moderate scenario		Optimistic scenario		Time	
Supply (m³/s)	Supply (m³/s)	Demand (m³/s)	Supply (m³/s)	Demand (m³/s)	Supply (m³/s)	Demand (m³/s)
20	20	40	30	48	20	01.01.2017
22	15	43	25	49	20	11.01.2017
25	18	63	29	60	20	22.01.2017
27	20	86	32	65	29	01.02.2017
32	22	70	37	80	30	11.02.2017
27	21	60	39	65	30	21.02.2017
33	25	72	40	83	35	02.03.2017
35	24	77	40	85	35	12.03.2017
39	38	88	69	95	45	23.03.2017
36	40	77	60	85	50	01.04.2017
38	40	80	69	89	45	11.04.2017
39	20	82	30	90	19	22.04.2017
40	15	85	22	93	20	01.05.2017
45	15	98	23	100	19	11.05.2017
62	25	130	40	155	33	22.05.2017
58	20	128	36	140	30	01.06.2017
55	23	120	40	135	30	11.06.2017
58	40	122	62	137	50	21.06.2017
59	41	125	62	140	50	01.07.2017
57	38	122	58	137	45	11.07.2017
59	130	129	200	145	145	21.07.2017
50	162	110	35	125	175	01.08.2017
52	188	112	270	127	205	11.08.2017
49	90	108	130	120	100	21.08.2017
40	112	88	165	95	125	31.08.2017
30	150	68	220	75	165	10.09.2017
30	200	67	283	73	210	19.09.2017
33	215	72	310	82	232	29.09.2017
32	215	70	310	80	232	09.10.2017
35	190	79	277	98	205	19.10.2017
30	160	77	35	75	175	29.10.2017
29	140	75	207	70	145	09.11.2017
22	109	50	160	55	115	18.11.2017
20	72	40	104	49	80	28.11.2017
19	30	39	47	45	35	08.12.2017
20	15	40	283	47	19	18.12.2017

The comparison among different scenarios proved that, the water shortage differs due to change in the implementation rates of different actions. These actions are: control of rice areas, application of modern irrigation techniques in new agricultural lands, enhancement of irrigation network efficiency, making successful public awareness campaigns for water use, and application of water saving technologies for municipal and industrial sectors. The water shortage will increase from 1.6 currently to 1.85 BCM/year in the pessimistic scenario, if implementation rates are low. However, the shortage will decrease to 0.59 BCM/year in the optimistic scenario without exceeding the maximum allowed drainage effluent to Quarun Lake (0.35 BCM/year), if the implementation rates are high. Water supply should be reduced for the period between January and June in the optimistic scenario to save the observable water excess.

Future water shortages for the Godavari river basin, India have also been obtained by RIBASIM model through the Hydrology Project, Water Resources Department Government of Maharashtra, India. The simulation results showed that the water requirement will increase from 2155 in the year 2000 to 4219 and 4768 Mm^3 by the years 2015 and 2020, respectively. The water shortage will increase from 234 in the year 2000 to 1908 and 2375 Mm^3 by the years 2015 and 2020, respectively. Results of Godavari river basin, India are in agreement with the current study. From the simulation results of the Godavari river basin, the following recommendations have been reported to minimize the water shortages:

- Increasing irrigation efficiency from 39 to 55% by modernization and improvement of existing conveyance;
- Efficient management of public water supply schemes by using closed pipe supply system and thereby reducing transit losses from 50% to 10%.

CONCLUSIONS AND RECOMMENDATIONS

The current water shortage in Fayoum Governorate is 1.6 BCM/year without reuse of drainage water and wastewater. According to RIBASIM, the water shortage will range between 0.59 and 1.85 BCM/year in optimistic and pessimistic scenario for the year 2017. This water shortage reduction is a result of high implementation rates of different actions. The actions are classified under supply management, and demand management. The supply side includes implementation of new projects that increase the Nile water supply such as New Bahr Kouta, Bahr Gerza, and Bahr Wahby projects. The demand side includes control of rice area and other crops having high rates of water consumption, application of modern irrigation techniques in new lands, enhancement of irrigation network efficiency, making successful public awareness campaigns for water use, and application of water saving technologies for municipal and industrial sectors. In the optimistic scenario, the water shortage and the drainage water volume were the lowest, although the agricultural area was the largest. The optimistic scenario was achieved, when the tested packages of actions are implemented with high rates. Since Fayoum is a miniature of Egypt with regard to natural and water resources systems, the results of this study can be used as guidelines for optimization of the water resources system in Egypt. The present study recommends the followings:

Water supply release at Lahon Dam should be reduced in the period between January and June to save 0.9 BCM of the excessive water, and hence to reduce the volume of drainage effluents and to keep the safe water level in Quarun Lake. This will protect the surrounding areas from the over-flooding.

New projects should be implemented to transfer the drainage water effluent away from Quarun Lake such as Al-Katea and Al-Tagen drain stations, which are planned to be installed by the Ministry of Water Resources and Irrigation, Egypt. The two stations can lift drainage water into Bahr El-Bashawat canal to keep the safe water level in Quarun Lake and irrigate new 50,000 Feddans.

According to the soil analysis, sprinkler and drip irrigation systems should be applied in the new lands. Since these lands have high infiltration rates and water percolates from them into agricultural lands in Fayoum depression.

REFERENCES

1. Arranz, R., McCartney, M.P., 2007. Application of the Water Evaluation and Planning (WEAP) model to assess future water demands and resources in the Olifants Catchment-IWMI Working Paper 116. IWMI – International Water Management Institute, South Africa, pp. 91.
2. Brouwer, C., Prins, K., Kay, M., Heibloem, M., 1988. Irrigation Water Management: Irrigation Methods. FAO – Food and Agriculture Organization of the United Nations, Training Manual No. 5. Available from: www.fao.org/docrep/S8684E/S8684E00.htm

3. Georgakakos, A.P., 2006. Decision support systems for integrated water resources management with an application to the Nile Basin. In: Castelletti, A., Soncini Sessa, R. (Eds.), Topics on System Analysis and Integrated Water Resources Management. Elsevier, Georgia, USA.
4. MWRI, 2001. Technical Report No. 19: Agricultural Sector Model for Egypt (ASME). MWRI – Ministry of Water Resources and Irrigation, Egypt.
5. NWRP/MWRI, 2013. NationalWaterResources Plan/Ministry ofWaterResources and Irrigation, Egypt. FinalReport: The NationalWaterResources Plan for Egypt – 2017.
6. Schulze, R.E., Smithers, J.C., 2004. The ACRU agrohydrological modeling system. In: Schulze, R.E. (Ed.), Modeling as a Tool in Integrated Water Resources Management: Conceptual Issues and Case Study Applications – WRC Report 749/1/04. Water Research Commission, Pretoria, South Africa, pp. 47–83.
7. Shendi, M.M., Khater, E.A., Gomaa, O.R.,2006. Land evaluation of some soils east of El-Fayoum Governorate adjacent to Assuit desert road. In: Agriculture and Food in Middle East, 3rd Egyptian-Syrian Conference. El-Minia University, Egypt, Available from: http://www.fayoum.edu.eg/Agriculture/SoilWater/pdf/LAND.pdf
8. Van derKrogt,W.,Verhaeghe,R., 2001.Regional allocation ofwaterresources.In:Hengsdijk,H.,Bindraban, P.(Eds.), Proceedings of an International Workshop on Water Saving Rice Production Systems. 2001 April 2–4. Plant Research International, Wageningen, Netherlands, pp. 101–116.
9. Van der Krogt, W., 2010. Technical Report: RIBASIM Version 7.01 User Manual Addendum. Deltares Institute, Netherlands.

CITATION

Mohie M. Omar, Evaluation of actions for better water supply and demand management in Fayoum, Egypt using RIBASIM, Water Science, Volume 27, Issue 54, October 2013, Pages 78-90, ISSN 1110-4929, http://dx.doi.org/10.1016/j.wsj.2013.12.008.

CHAPTER 9

Development and Uptake of Scenarios to Support Water Resources Planning, Development and Management – Examples from South Africa

Nikki Funke[1], Marius Claassen[1] and Shanna Nienaber[1]

[1]Natural Resources and Environment Unit, Council for Scientific and Industrial Research, Pretoria, South Africa

INTRODUCTION

The international agenda on water resources development reflects societal needs, political agendas, economic realities and the state of resources. The industrial revolution, which started in the 18th century, brought social and economic prosperity but also marked a major shift in humanity's impact on the earth's systems. This shift is now referred to as the Anthropocene [1], where humans have brought such vast and unprecedented changes to the planet that this era represents a new geological time interval [2]. Societal needs have shifted since the 1940s from a need for modest food production to a need for increased agricultural productivity that has been met by high yield crops, the use of pesticides, the application of fertiliser and advanced agricultural techniques. This development has averted food shortages, but has also resulted in humanity having to pay a heavy price in terms of increased water use and energy consumption, as well as environmental degradation [3].

From the early 1970s a series of events and key documents has promoted an integrated approach to sustainable development. The 1972 United Nations Conference on the Human Environment considered the need for a common outlook towards the preservation and enhancement of the human environment [4]. The World Commission on Environment and Development advanced this agenda in their report 'Our Common Future', with an emphasis on sustainable development promoting harmony among human beings and between humanity and nature [5]. The International Conference on Water and the Environment that took place in Dublin in 1992 resulted in the development of four guiding principles [6]. These principles, commonly referred to as the Dublin principles, state that: water is a finite resource with economic value and social implications; local communities must participate in water management; water resources management must be developed within a set of policies; and the role of rural populations and women should be recognised. This led to the Rio Declaration and the adoption of Agenda 21, which is a comprehensive plan of action to be implemented globally, nationally and locally in every area in which humanity impacts on the environment [7]. This declaration subsequently became the blueprint for sustainable development worldwide [8].

Uncertainties about societal, economic, political and environmental aspects have proved to be a considerable obstacle to the implementation of sustainable development. Here follow a few examples of such uncertainties. In 1980, the World Development Report of the United Nations [9] estimated that the world population would reach 6.029bn by the year 2000. Five years later, the estimate was updated to 6.088bn [10], with further updates at five yearly increments resulting in estimates of 6.194bn and 6.123bn [11, 12]. The actual population in the year 2000 turned out to be 6.188bn [13]. Future economic development is also uncertain, with the annual growth in Figure 1 showing how the world average varies significantly between years and also how the growth of individual countries (South Africa in this case), does not necessarily follow the global trend and is even more variable between years.

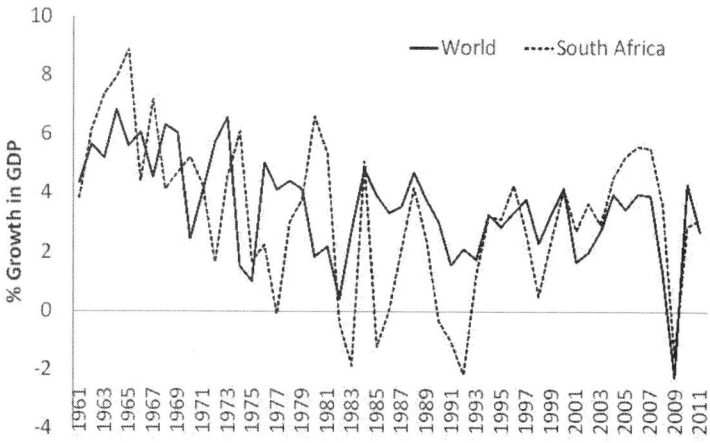

Figure 1. Annual economic growth between 1961 and 2011 [13]

INTRODUCTION

Environmental conditions also vary significantly over time and space, with Figure 2 illustrating the annual deviation of rainfall over southern Africa. This uncertainty is exacerbated by climate projections, which suggest that freshwater resources are vulnerable and have the potential to be strongly impacted by climate change, with wide-ranging consequences for human societies and ecosystems [14].

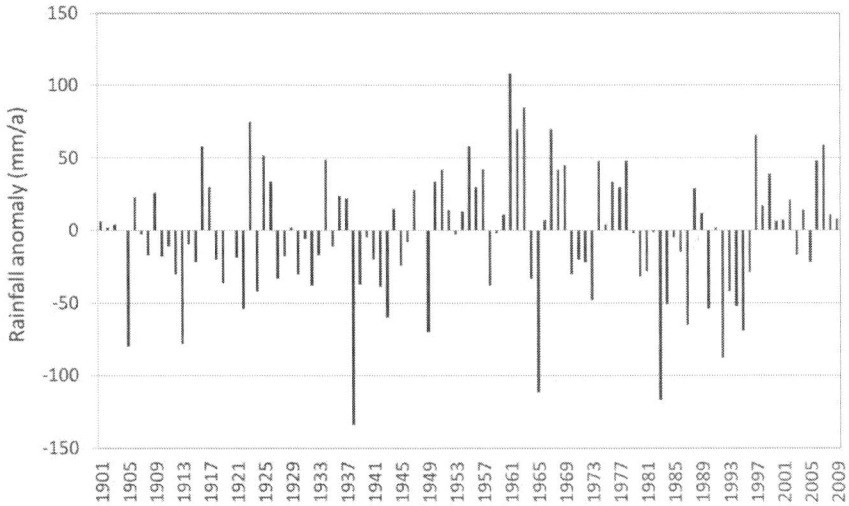

Figure 2. Annual rainfall anomalies for the southern African region (1901-2009; Adapted from [15]).

A question that emanates from the realisation that we live in a changing world where change is unpredictable is, 'How do we plan for the future?'
Water use in South Africa was first regulated through the Irrigation and Conservation of Waters Act (Act No. 8 of 1912), which managed the use of water from public streams for domestic, irrigation and industrial purposes [16]. The Water Act (Act No. 54 of 1956) further regulated water use by providing for the control of water pollution and the more effective protection of water resources. The variable distribution of water required the development of infrastructure to capture, store and distribute water. The subsequent expansion of mines, industries and urban areas created a demand for further infrastructure development. When this demand further increased and the social and economic issues in South Africa became increasingly complex in the 1990s as the country was transitioning from apartheidto democracy, a shift in thinking was required. As it became clear

that engineering solutions to increase water supply were not sustainable, a holistic strategy to meet future needs became more popular [17].

The new National Water Act (Act No. 36 of 1998) [18] emphasised water resources management at national and catchment scales, made specific provisions for the protection of water resources, established mechanisms to ensure equitable and efficient water use and promoted participatory management. The National Water Resources Strategy [19] addressed the balance of future water supply and demand by establishing scenarios. The demand scenarios were based on population growth by 2025, with the high population scenario at 54 million people and the low population scenario at 50 million people. It also established economic growth scenarios, with the upper scenario assuming 4% growth in GDP and the less favourable scenario assuming 1.5% [19].

While there has been much progress in water infrastructure development for services (public benefit), the backlog in issuing water use licenses (mostly for private benefit) stood at 4 318 in 2011 [20]. The protection of water resources has suffered as a result of the government's drive to achieve social and economic development, with South Africa ranked 128 out of 132 countries in the Environmental Performance Index [21]. The National Water Act provides for a balance of responsibilities, ranging from the Minister and Director General at the national level, to Catchment Management Agencies (CMAs) at the basin level and Water User Associations (WUAs) at a sub-basin level. Progress has been slow as after 14 years after the promulgation of the Water Act, only two CMAs (out of the 19 intended) have been established [20]. It can be argued that many hurdles have to be overcome to fully realise cooperative governance for Integrated Water Resources Management (IWRM), with inadequate human and institutional capacity being one of the main factors limiting the efficient management of water resources in South Africa [22]. To illustrate this point: the country's Department of Water Affairs (DWA) reported having 4 286 people in its employment in September 2010, while 1 155 posts were vacant at the time [20].

From the discussion above it becomes clear that we live in a world with social, economic and environmental conditions that are variable and difficult to predict, and the water sector is no exception. This uncertainty provides a challenging environment for policy and institutional development. Scenarios are one way of attempting to achieve a desired outcome in an uncertain and variable future [23]. The rest of this chapter will examine the research question, 'How are scenarios able to achieve impact in an uncertain world, with a particular focus on water resources planning, development and management?' The body of this chapter focuses on the research method, presents an overview of scenario development and the importance of scenario development and how they facilitate more effective water resources planning, development and management, focuses on a few select South African scenarios and the impact they have had and then turns to discussing the impact of scenarios in general. The conclusion wraps up the learning from this chapter and suggests a way forward in terms of future research and designing scenarios for impact.

THE ABILITY OF SCENARIOS TO ACHIEVE IMPACT IN AN UNCERTAIN WORLD WITH A FOCUS ON WATER PLANNING, DEVELOPMENT AND MANAGEMENT

The authors of this chapter conducted an exploratory study on the ability of scenarios to achieve impact in an uncertain world, with particular reference to water planning, development and management. They conducted a review of scenario planning literature in the water and other sectors, and also considered literature focusing specifically on the impact of scenarios. The authors also considered literature on the impact of scientific research and on the science-policy interface. This was accompanied by a search of major databases (e.g. Google Scholar, EBSCO Host and Scopus) to determine where and how the four scenarios discussed in this chapter have been cited. In addition, the authors interviewed selected stakeholders in the water and other sectors who are likely to have been exposed to scenarios and who may use scenarios when making decisions in their workplace.

Scenarios and Their Importance in the Water Sector

The concept of scenario planning has its origin in military applications, with the US Air Force developing 'scenarios' of what the enemy might do and preparing alternative strategies. It was thus aimed at achieving a desired outcome in an uncertain future [24]. At the end of the 1940s, researchers at the RAND Corporation started to investigate the scientific use of expert opinion in planning for the future [25]. The Royal Dutch Shell company employed scenario tools to good effect in the 1970s, when they improved their size and profitability by being prepared to act quickly during the oil price shock of 1973 [26]. In summarising definitions of scenarios, scenarios can be described as a narrative description of a possible state of affairs or development over time, that they are useful to communicate speculations about the future to promote discussion and feedback, and that they can dramatise trends and alternatives, explore the impacts and implications of decisions, choices and policies, and provide cause-and-effect explanations [24].

Clem Sunter is credited with popularising the use of scenarios in South Africa, with 'The World and South Africa in the 1990s', which describe the 'High Road' and 'Low Road' scenarios [27]. The publication was based on work from Anglo American Corporation teams in London and Johannesburg. Subsequently, Adam Kahane facilitated a process that became known as the Mont Fleur scenario project, which was launched in 1992. It explored the question of 'What will South Africa be like in the year 2002?' These scenarios were arrived at collaboratively by a very broad group [28]. The Department of Arts, Culture, Science and Technology (DACST) also deployed scenarios and technology foresighting in the development of South Africa's National Research and Development Strategy, with Kahn initiating and leading the development of the South African National Research and Technology Foresight Project [29]. The Dinokeng team [30] developed '3 Futures for South Africa', which characterised future scenarios based on the effectiveness of the state and the engagement of society. Some of the recent scenario projects in the water sector include the World Business Council for Sustainable Development report on 'Business in the World of Water: WBCSD Water Scenarios to 2025' [31], and the Global Research Alliance (GRA) report on 'Science and Technology-based Water Scenarios for sub-Saharan Africa' [32].

The Importance of the Use of Scenarios in Water Resources Planning, Development and Management

Scenarios are important and useful to water resources planning, development and management in a number of ways. In the South Africa context, in particular, scenario development processes have been instrumental in initiating strategic conversations among scenario workshop In the South African context, (e.g. the transition from apartheid to democracy), and have helped develop a common language among people with widely divergent views [28]. Those involved in scenario development processes may be inspired to think more broadly about the future and the forces creating it. They may also realise how their particular actions may help to create a desired future [33]; and they may have suggestions about which options exist to direct target audiences on to a desirable path [28]. The knowledge that scenarios generate can therefore potentially empower role players in the water sector and other sectors to engage in participative governance by equipping them with insights into potential futures they may face, and making them aware of the implications of certain decisions, behaviours and actions [23]. Finally, the advantage of communicating scenarios as stories is that they have the psychological impact that other more academic means of communication, for example, graphs and equations, lack. Stories give order and meaning to events, which is crucial for imagining future possibilities [34].

Some South African Scenarios: Overview and Impact

The discussion in this chapter and the research question were inspired by the development of the Water Sector Institutional Landscape by 2025 scenarios. These scenarios were the main output of a research project led by the authors. In particular, the authors are interested in how these scenarios could be used by potential end-users. Given this question and the importance and potential usefulness of scenarios in facilitating decision-making in a context of uncertainty, it becomes important to reflect on some examples of scenarios that have been developed in South Africa at different points in history and to learn from the impact they have had on different sectors, including the water sector. These scenarios are discussed in chronological order. The section starts with the High Road/Low Road scenarios that were developed late in the apartheid era and on the cusp of South Africa's transition to democracy. Secondly, the Mont Fleur scenarios, which were developed during the democratic negotiations, are discussed. Thirdly, the section focuses on the Dinokeng scenarios that were developed in 2009, the year a new president came to power and a serious economic crisis shook the world. The section concludes with the Water Sector Institutional Landscape scenarios that focus on potential futures of the South African water sector in 2025.

Overview and Process

The High Road/Low Road scenarios were an initiative by the Anglo American Corporation in the early 1980s and aimed to look into some less conventional approaches to business planning and future investment decisions, given the international economic turbulence of the 1970s and the resultant slump in commodity markets. During this time, South Africa's economic performance was poor and several events resulted in the country becoming increasingly isolated and the government resorting to a rule of force. Careful and gradual reforms by the apartheid government in the middle to late 1980s and increasing attempts by members of the white establishment to reach out to black leaders in exile, led to the eventual unbanning of the African National Congress (ANC) and the release of Nelson Mandela in 1990 [35].

The scenarios involved a large-scale exercise with numerous contributors, notably Pierre Wack and Ted Newland, as well as Clem Sunter. Most of the effort went into developing global scenarios which were based on the analysis of key 'drivers' (for example, demography, technology and societal values) of developments in Japan, the USA and USSR (then regarded as the main players of the world economy), and also the ingredients for success of 'winning' nations and world class companies. This work then provided the basis for the South African scenarios. In essence, these scenarios focus on the choice the country was facing to either (through consultation and negotiation) travel on the 'High Road' to a non-racial democracy and increasing prosperity, or, to continue on the 'Low Road' of confrontation, conflict and falling incomes (as a repressive, centralised society and controlled economy) and ending up as a 'waste land' [35].

The scenarios conclude with the need for a 'common vision' to help launch South Africa into the more desirable 'High Road' scenario. This common vision entails putting South Africans first (looking beyond different races and groups), to turn the country into a 'winning' nation and to work towards achieving a certain income per head, all of which would be reached through negotiation [28].

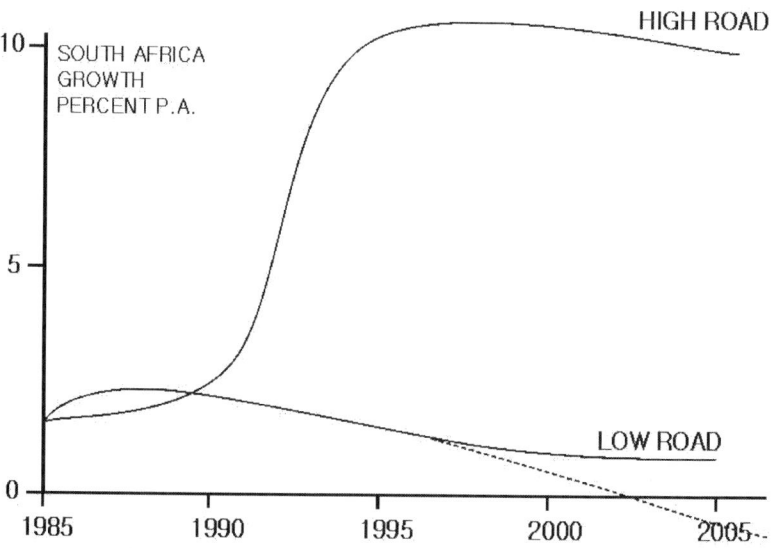

Figure 3. The High Road/Low Road scenarios depicting two possible future trajectories for South Africa [27]

THE ABILITY OF SCENARIOS TO ACHIEVE IMPACT IN AN UNCERTAIN WORLD WITH A FOCUS ON WATER PLANNING, DEVELOPMENT AND MANAGEMENT

Dissemination and Impact

Within a year, starting in 1986, Clem Sunter presented the 'High Road/Low Road' scenarios to 230 (mostly white) audiences at various levels of society, thereby reaching between 25 000 to 30 000 people [35]. Senior politicians of the ANC were also one of Sunter's audiences before the eventual negotiated settlement was reached [28]. The message of the scenarios seems to have made a big impression on the audiences as it was ultimately positive and encouraged people in the country to take it into their own hands to get on to the 'High Road', without being prescriptive about how this should be done [35]. In particular, the High Road/Low Road scenarios also seem to have contributed somewhat to the shift in thinking in government circles, and indeed as supporting evidence for a need for change, which eventually brought about a political transition. In conclusion then, the High Road/Low Road scenarios started out as a corporate scenario project and resulted in a brilliant communication exercise, both in terms of content and style of presentation, that reached thousands beyond the initial intended audience and paved the way for more prominent South Africa scenario exercises to come [35].

In terms of uptake in the scientific and decision-making community, Clem Sunter's book 'South Africa and the World in the 1990s' has been widely cited and includes discussions of a range of topics. These include reflections on various elements of the political and economic transformation of South Africa, the future of Africa, scenario development and planning and globalisation. The citations include a variety of different sources, including books, journal articles, theses and reports. These sources are mostly from the economic, management and social sciences, but also from the health and environmental sciences.

While no examples could be found of the use of the High Road/Low Road scenarios in the water sector, it is likely, judging from the fact that Sunter presented these scenarios to such a wide range of audiences, that some members of government and other stakeholders in the water sector would have been exposed to them in the late 1980s or early 1990s. South Africa's new water legislation certainly reflects the thinking associated with the High Road scenario, with emphasis on introducing ground-breaking new principles into the governance of South Africa's water resources. Though somewhat outdated now, the High Road/Low Road scenarios serve as a reminder of where South Africa could be headed at any point in history. In terms of water resources, South Africa is in need of thoughtful planning, development and management if its water resources are to continue to meet the needs of its ever growing and developing population.

Overview and Process

The Mont Fleur scenarios were developed in South Africa between 1990 and 1994. Key events during this time were the release of Nelson Mandela, and the legalisation of the ANC, Pan African Congress (PAC) and South African Communist Party (SACP) [36]. The country's first racially inclusive elections were also held at this time. Given this political climate, multiple forums emerged that brought a broad range of stakeholders together to try to develop a new way forward for South Africa. In particular, issues such as housing, education, and constitutional reform received attention [35, 36].

The Mont Fleur scenarios formed a part of this process and essentially tried to encourage debate, thinking and imaginative ideas around how to shape the first ten years of the 'new' South Africa and also to illustrate how certain choices would steer the country towards different outcomes. The Mont Fleur scenario team was made up of a diverse group of 22 prominent South Africans, including politicians, activists, academics and business people [36].

The Ostrich scenario represents a continuation of the status quo in South Africa and suggests that no negotiated settlement would be reached and that government would continue to be non-representative [37].

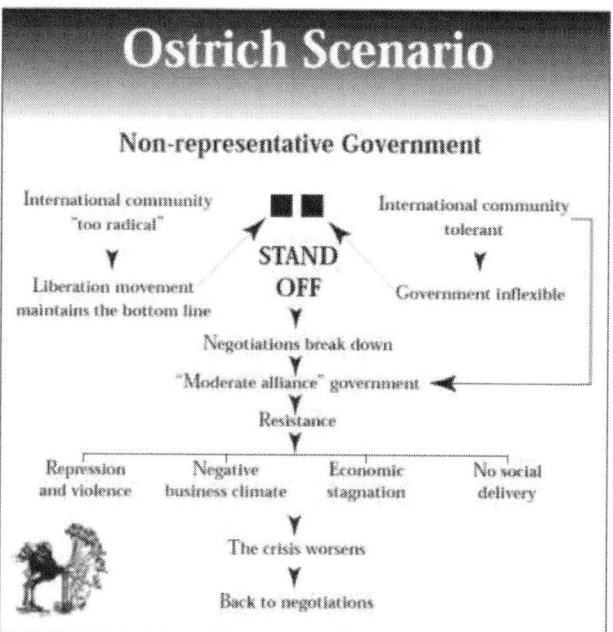

Figure 4. The Ostrich scenario [36]

THE ABILITY OF SCENARIOS TO ACHIEVE IMPACT IN AN UNCERTAIN WORLD WITH A FOCUS ON WATER PLANNING, DEVELOPMENT AND MANAGEMENT

The Lame Duck scenario suggests a South Africa where a settlement would have been achieved but where the transition to a new dispensation would be slow and indecisive [37].

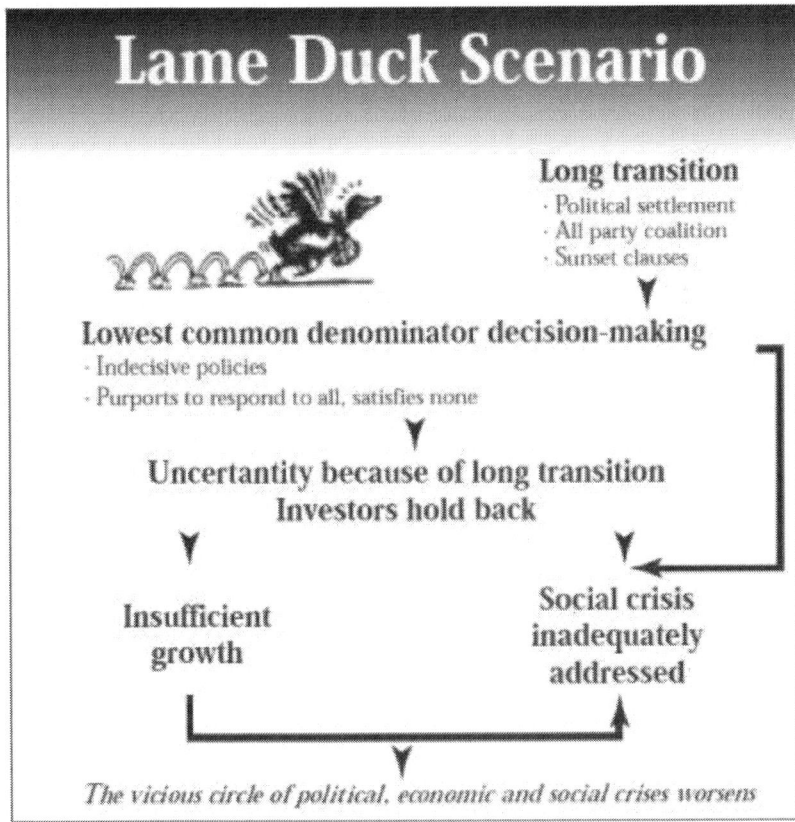

Figure 5. The Lame Duck scenario [36]

The Icarus scenario suggests a rapid transition to a new government that would push for populist and unsustainable economic policies [37].

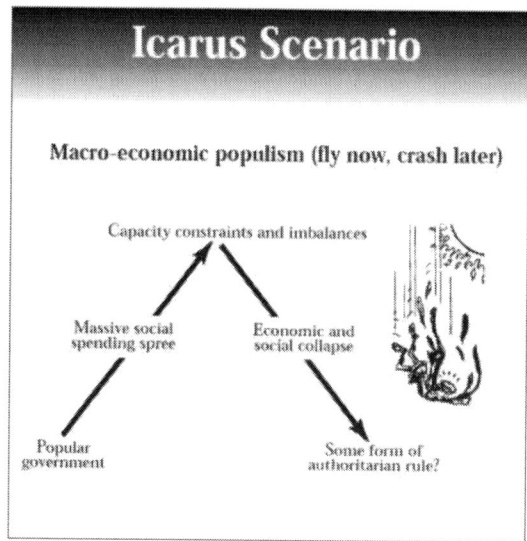

Figure 6. The Icarus scenario [36]

The Flight of the Flamingos scenario depicts a government that would choose sustainable policies that would lead the country towards inclusive growth and a maturing democracy [35, 36, 37].

Figure 7. The Flight of the Flamingos scenario [36]

THE ABILITY OF SCENARIOS TO ACHIEVE IMPACT IN AN UNCERTAIN WORLD WITH A FOCUS ON WATER PLANNING, DEVELOPMENT AND MANAGEMENT

By means of a process of negotiation and reflection on different drivers and concerns, the Mont Fleur scenario team was able to articulate a range of potential outcomes for South Africa during the 1992 to 2002 period. This also helped to clarify the goals and aspirations related to where the country should be heading.

Dissemination and Impact

A variety of dissemination techniques were used by the Mont Fleur team. Key to this process was the fact that each of the individual participants took responsibility for spreading the message of these scenarios. They did this by presenting and discussing the scenarios with more than 50 different groups of people including political parties, companies, academics, trade unions and civil society organisations [36]. This was possible given the diverse background that the team came from. Over and above this process, the scenarios were condensed into an easily accessible 14 page document. This document was distributed to national newspapers. A short video was also produced that combined cartoons with presentations by team members [35, 36].

The impacts of this scenario development process are subtle. Three key points are important in this regard. Firstly, Mont Fleur, along with other processes taking place in South Africa at the time, helped to establish a common language and understanding about the challenges facing the country and the way forward. This was because participants focused on an issue of common concern for all: 'the future of South Africa'. Secondly, although participants could not agree on one major solution to South Africa's problems, they could agree that certain solutions would not work (such as armed revolutions, continued minority rule and socialism). Thirdly, through an informal process of open conversation, participants who had not expected to agree with each other found common ground and shared understandings about the future of the country [36]. Given these points it is clear that the impact that the Mont Fleur scenarios had was first and foremost on the individuals who participated in the process. There was subsequently a more indirect impact on broader society once these individuals started presenting the scenarios to their various constituencies. Given the widely publicised nature of South Africa's political transition, these scenarios also gained popularity overseas [35].

The Mont Fleur scenarios have also been cited in a range of publications. These citations occur in journals, books, conference papers, dissertations and magazines that focus on a range of different disciplines, namely the social, natural and technical sciences. Given this broad interest, the publications cover a broad range of topics most of which are geared towards futures research, democratic transition and strategic planning. This citation record illustrates that the Mont Fleur scenarios seem to have had a considerable impact on the academic community.

Whilst the Mont Fleur scenarios are not obviously related to the South Africa water sector, they did contribute to setting a precedent for using scenario development for planning purposes in South Africa. So, for instance, as mentioned above, the National Water Resources Strategy established a set of water demand scenarios. As with the High Road/Low Road scenarios, the Mont Fleur scenarios were part of the thinking and move towards democratic transition in South Africa. As a result of and in order to complement this change, the water sector was fundamentally transformed and restructured.

Dinokeng Scenarios

Overview and Process
The Dinokeng scenario team consisted of 35 leaders from civil society, government, business, political parties, public administration, trade unions, religious groups, academia and the media. The scenario development process was sponsored by the financial institutions Old Mutual and Nedbank who believed that, 15 years into South Africa's democracy, it was important to initiate a reflective and constructive debate about the country's future. According to the Dinokeng scenario team, some of the most prominent challenges facing South Africa are unemployment and poverty, safety and security, education and health. These challenges appear all the more grave in the context of a volatile global economic market, and a global economic crisis that shook the world when these scenarios were developed in 2009 [30].

The Dinokeng scenario team agreed that South Africa needs to realise that the country has failed to appreciate or understand the imperatives of running a modern democratic state, and that there is a problem with the country's self-interested, unethical and unaccountable leadership across all sectors. Additional problems include a weak state that is increasingly less capable of addressing the country's critical challenges, and a population that is either not interested and is showing a growing dependence on the state to provide for everything, or has become co-opted into government or party structures since 1994 [30].

The scenario team developed three possible scenarios which the country could be heading into:

Firstly, the Walk Apart scenario suggests the state becoming increasingly weak and ineffective, and the population, which is looking out for its own interests, eventually losing patience with the state and resorting to protest and unrest to make its views heard. Because the state is unable to meet the population's demands and expectations, it responds brutally, and the result is a spiral of resistance and repression. The Walk Apart scenario therefore suggests a need for South Africans to address their critical challenges, to build state capacity and to organise themselves to engage government in a constructive way, in order to prevent themselves from heading towards disintegration and decline [30].

Secondly, the Walk Behind scenario suggests the state becoming increasingly confident and strong in terms of leading and directing development, fuelled by the fact that civil society is becoming more and more dependent and compliant. The problem is that the state does not have the capacity to address the critical challenges the country is facing on its own. The message of this scenario is that state-led development cannot be successful if there is insufficient state capacity. Furthermore, if the state intervenes constantly and dominates all other sectors, it will crowd out private business and civil society initiatives and will end up creating a population that is complacent and dependent on the state [30].

Thirdly, the Walk Together scenario suggests the state becoming collaborative and increasingly listening to its citizens and leaders from different sectors, engaging with critical voices, and consulting and sharing authority in order to work towards long-term sustainability. In this scenario there is also a focus on a population that takes leadership and holds government accountable and shows an active interest in policy development and outcomes. It is important that South Africans re-engage, that the capacity of the state is strengthened and that leaders from all sectors think beyond their own self-interest and contribute to nation-building [30].

In conclusion, the present contains the seeds for all three futures to be realised. For a healthy democracy and strong socio-economic development to persist, it is important to have a healthy interface between an effective state and an alert and involved population; the nature of this interface is likely to determine the future of the country [30].

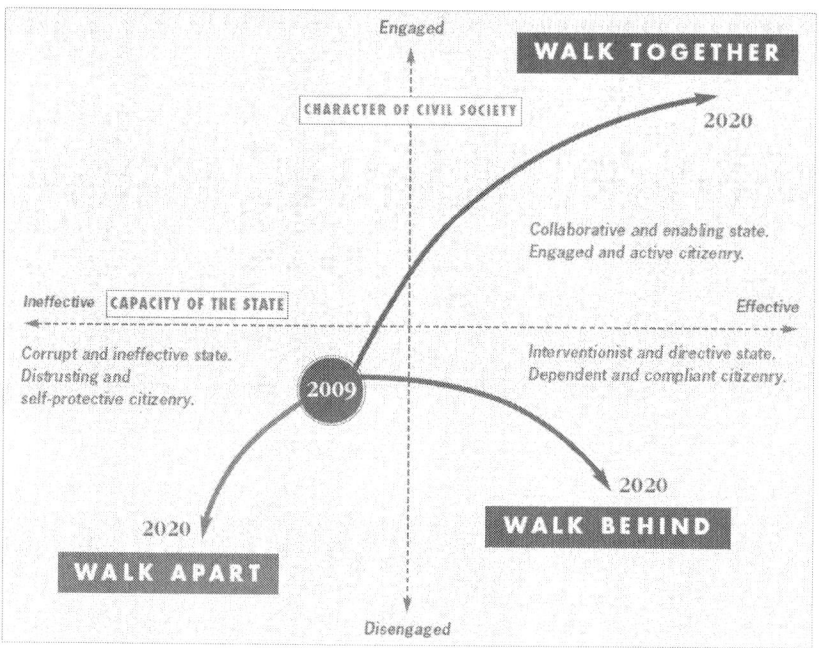

Figure 8. The Dinokeng scenarios [30]

Dissemination and Impact

In terms of dissemination and impact, once the Dinokeng scenarios on possible futures for South Africa had been developed, the messages of these scenarios were shared with a range of stakeholders. This engagement was followed up with a media and engagement campaign to communicate the Dinokeng scenarios to a variety of organisations, groups and communities across South Africa [30]. The Dinokeng scenarios and the process around their development were also placed on the Dinokeng scenarios website, which is a user-friendly resource for those who are interested in finding out more about these scenarios. The Dinokeng scenarios text is also available for download here.

A database search showed that these scenarios have been cited in a wide range of publications. These publications include discussion papers, theses, conference presentations, books and journal papers. The topics of the publications that cited the Dinokeng scenarios are wide-ranging and include issues around interrogating and addressing social issues related to South Africa's democracy, such as local government, education, housing, poverty, unemployment and food security. Many of these topics have a future-centred focus, e.g. investigating South Africans' perceptions about the future, or planning for the future in local government structures. The fact that the Dinokeng scenarios were cited in different kinds of publications and across different subject matters indicates that, at least among the research community, the scenarios were widely distributed and taken up by researchers from different social science-based backgrounds and interests.

A question that arises here is to what extent the Dinokeng scenarios may be of relevance to water resources planning, development and management. While no examples of their use in the water sector were found, it can be argued that the insights provided by these scenarios would prove valuable in focusing on resolving some of the water governance related issues South Africa is currently facing. Examples include problems around water pollution resulting from ineffective waste water treatment and mine and industrial effluents, and water service delivery to previously disadvantaged communities. Those who need to address these water governance related problems could benefit from taking into account the need for maintaining a balance between strong and effective leadership in all sectors and an interested and engaged population, and reflecting on the different future directions such a relationship or lack thereof could take.

Overview and Process

An example of scenario development with particular reference to the South African water sector is the South African Water Research Commission's (WRC) Water Sector Institutional Landscape by 2025 scenarios, developed by the authors in 2011 with the assistance of Chan tell Illbury as facilitator, and in consultation with a range of water sector related experts and stakeholders. The focus of the scenario development was on water resources management in South Africa, also with relevance to the water services sector. The aim of these scenarios was to build knowledge about key drivers and uncertainties that relate to the future of the South African water sector, and specifically about the context in which water institutions may operate in future [23].

The knowledge for this project was generated through a structured research process to target existing and new institutional structures and to ensure the involvement and participation of a broad range of stakeholders. The aim of this engagement was to identify water-related needs, priorities and uncertainties based on a wide range of perspectives. A broad range of methods was employed to include stakeholders from both rural and urban environments and with different cultures and educational backgrounds. These included interactive workshops, semi-structured interviews, and a web-based survey. This process was characterised by continuous assessment, learning and adaptation [23].

The key drivers and uncertainties that were identified were subsequently translated into different scenarios that hold potential implications for social and economic development, as well as water resources and services in South Africa. The four scenarios were derived from a matrix with two axes that represent the ability of the decision-making paradigm of water institutions to deal with 'complexity' (refer to the x-axis of the diagram), and the reconciliation of environmental, social and economic demands of present and future generations (referred to as 'sustainability' on the vertical or y-axis of the diagram) [23].

Four possible scenarios emerged from the matrix. The Greedy Jackal scenario depicts a South Africa where water is scarce but government still struggles to meet developmental demands and address backlogs. Under these urgent socio-economic circumstances, environmental responsibility is not prioritised. Despite this the need for a multidisciplinary response to complex water challenges is acknowledged [38].

The Wise Tortoise scenario suggests that a paradigm shift has occurred resulting in a water sector that is multi-layered and engages many different sectors given the strategic importance of the resource in all facets of development. This approach allows for proactive management rather than crisis response to challenges [38].

The Busy Bee scenario suggests that the water sector is defined by great intentions but does not follow up on these with necessary actions. Thus, whilst rhetoric embraces sustainability, in practice there is limited economic and social development to support this process. Part of the challenge is a lack of civil society engagement and failure to embrace the complexity facing water resources management [38].

The Ignorant Ostrich scenario suggests that government fails to recognise water as central to development. As such they rush to implement politically appealing but imbalanced and short term solutions. Civil society is not engaged in decision-making and the complexity inherent to the water sector is overlooked [38].

THE ABILITY OF SCENARIOS TO ACHIEVE IMPACT IN AN UNCERTAIN WORLD WITH A FOCUS ON WATER PLANNING, DEVELOPMENT AND MANAGEMENT

Figure 9. The Water sector institutional landscape by 2025 scenarios [38]

Dissemination and Impact

The scenarios were printed by the WRC in the form of a colourful booklet and subsequently have been disseminated to some stakeholders. The scenario document and technical report documenting the scenario development process are also available online. While much more could have been done in terms of dissemination, this was not a component required by the project's funders and was therefore not planned into the project process from the start of the project. Therefore no funding was available to carry out this important part of the scenario development process. Nonetheless, these scenarios have the potential of feeding into the decision-making processes of water resources managers and decision-makers, but could also potentially empower a range of other role players in the water sector to engage in participative governance [23].

By studying the dissemination, impact and lessons learned from the South African scenarios discussed above, along with other literature related to the impact of scientific research and the science-policy interface, it is possible to distil some lessons and challenges relating to impact and how to more effectively produce and disseminate impactful scenario products.

A discussion on the impact of scenarios in general and reflections on such impact follows below.

The Impact of Scenarios

The previous section explored a number of South African scenarios in terms of their contents and impact. In terms of impact, Chantell Illbury and Clem Sunter refer to the "Wack" test, based on the ideas of Pierre Wack, a key scenarios planner in the 1970s and 1980s. According to this test, scenarios are not deemed important because of their prediction capability. What is important is their ability to influence the mindsets of decision-makers and to encourage them to act [39].

The issue of scenario impact is in many ways tied to a broader issue often referred to as the science-policy or science-end user interface. This issue essentially speaks to the challenge of getting knowledge that is produced by scientific or expert teams to be used in the public domain. This discourse recognises that there should be a close relationship between science or research products and their end-users, which could include government, policy-makers, businesses and communities. In reality, however, this relationship is not always an effective one, resulting in research often (or mostly) having minimal impact on policy and practice. The science-policy interface discourse explores why this happens in order to try to advise scientists and end-users about how to more effectively incorporate research into practice [40, 41, 42].

In terms of scenarios there tend to be two major opportunities for impact. The first is an impact on the participants who are part of the scenario development process. This is referred to as 'communication forscenarios' [43]. Similarly this opportunity for impact can be referred to as first order influence. First order influence refers to participants in the scenario development process undergoing personal changes in their thinking and behaviour. They also commit to the process, learn new skills, and build new networks and relationships. Because participants increasingly respect, understand and trust each other, they jointly commit to change [37].

The second is the impact of scenarios on broader society. This can be referred to as 'communication ofscenarios' [43]. Here a wider group of stakeholders ideally need to be exposed to the scenarios once they are fully developed. As such, at this stage it is important to think about ways to foster appropriate dissemination and use of scenarios. This stage can also be referred to as second and third order influence. Second order influence is closely linked to first order influence. Participants who have been part of the scenario development process go back to their communities and networks and start sharing their new language, thoughts and insights with others. Third order influence is a process of social change, but can be difficult to monitor and study because of the many variable factors that influence every change process [37].

The following sub-sections reflect on the impact of scenario development on the participating team as well as the impact or influence of scenarios on broader society.

Impacts on Participants in the Scenario Building Process
Participants in a scenario development process actively engage and transform the process in the sense that they are asked to share their views, ideas, concerns and experiences in order to generate drivers to develop scenarios or stories from these drivers. It is important to recognise that this kind of individual impact is difficult to quantify and tends to be very subtle [35; 37]. Nonetheless, the kinds of impacts that individuals experience can include:

- Experiencing reframed mental models – By being forced, through the scenario development process, to articulate and share different perspectives and mental models, participants are made to think carefully about their perceptions and often re-think their views when faced with other participants' views and the need to move collectively towards a desired future [44].
- Gaining a broadened network of relationships – Scenario development processes bring together groups of people to have open and constructive conversations. This process fosters a shared understanding, trust and a sense of community [44].
- Regenerating energy, commitment, and action – By clarifying desired futures and building consensus about how different actions will navigate society towards certain scenarios, a sense of regenerated energy and commitment can be achieved. Also, with new commitment in place, new actions can be catalysed [44].
- Taking pride in participation – When interviewed, participants tend to be quite proud of their involvement in scenario development processes. This encourages them to use and share the learning from the scenario development process during other projects and/or engagements [35].
- Creating a common vocabulary, trust and mutual understanding – Through the process of developing scenarios these subtle processes tend to be fostered. This is important as it is through trust and

understanding that people are able to work together towards a desired shared future [37].

Facilitating and Forming a Scenario Team

Whilst it is clear that a subtle process of impact and transformation can occur in a scenario team, this does not happen automatically. There are a number of lessons that have been learned through scenario development processes over the years that need to be borne in mind.

Firstly, having a diverse team is important [45]. The team should come from different age, race and gender brackets as well as a wide range of ideological spectrums [35]. This diversity is important because the more diverse the team is, the more diverse the driver inputs will be and as such the richer and more accurate the scenario development process will be. Also, an inclusive rather than exclusive scenario development process lends legitimacy to the process [46].

Secondly, embracing transdisciplinarity in any scenario development process is important. This implies that in order for scenarios to have the impact they need, they should be produced by a team made up of multiple different actors from government, civil society, communities, and research institutions. This will help the team to take into account different types of knowledge that different actors have (such as technical, traditional, experiential, cultural, and political knowledge). In so doing the inherent complexity in future planning processes will be reflected [47].

Finally, working with a diverse team with different knowledge, experience and viewpoints is not always easy. Conflict can arise when participants with different viewpoints are made to work together. Also, meaningfully incorporating feedback from diverse sets of stakeholders tends to be a highly time consuming process. Given these and other challenges that can arise, the importance of having a skilled, sensitive and insightful facilitator cannot be underestimated. Such a facilitator needs to be able to manage strong individuals who dominate conversation with their own agendas, and needs to be able to encourage everyone to express their opinions during the scenario development process [35].

The impacts of scenarios on broader society are harder to ascertain and measure than the impacts of scenario development on the scenario team itself. This is because there are no measurable criteria for quantifying the impact that scenario products have on society, be they in written or oral form. Also the outcome of scenario development processes can never be attributed to a single factor. Scenario development processes typically deal with broad developmental issues making the range of issues and actors that they try to affect diverse. Scenario development processes also happen within the context of a range of related social activities, such as developments in policy, civil society events and public debates. For example, in the case of developing the South African scenarios of the 1980s and 1990s, there were multiple social forums, political parties, and government groups working on transforming the country. These scenarios and their related processes were just one input amongst many others that were part of the broad transition process. Similarly, the Water Sector Institutional Landscape by 2025 scenarios exist alongside scenarios established by the National Water Resources Strategy, the various government departments that do strategic planning and forecasts in relation to water, and the host of grassroots organisations that work on managing water sustainably for the future. Any impact or change in the water sector must then be attributed to a whole range of interlocking factors rather than just one set of scenarios.

Facilitating the Effective Dissemination of Scenario Products to Society

In order for scenarios to have influence in the broader public space a number of key lessons are important. Firstly, a broad and extensive communication process is a key requirement and should be planned and budgeted for from the beginning of the project [42]. It is important that such a process targets multiple different actors in society, and takes place at many levels of scale (local, provincial, national) [23, 41, 42] in order to engage society and attempt to create a better future [37]. Non-government actors are an important target audience because they are critical in terms of instigating social debate, bringing about grassroots changes and challenging authorities to improve their performance [42].

In government, actors need to be aware of scenario products and how they can make use of them [23]. With regard to the South African water sector in particular, there seems to be a need to enable officials from DWA to apply the outcome of scenarios thinking and processes in their strategic decision-making aimed at mapping out the future of the water sector. A possible way of enabling experts and government officials to think imaginatively and creatively about the future, given their considerable daily workload and challenges, would be to involve scenario experts as facilitators for strategic planning sessions. Such sessions should ideally take the form of one or two day workshops in order to remove government officials from their immediate working environment and enable them to apply their minds to thinking creatively and focusing exclusively on the planning task at hand [23]. When engaging with government departments, it is important to be sensitive to and aware of different issues inherent in the government hierarchy. Non-political, technical experts tend to have a good knowledge of technical issues, but it is also important to target more senior political actors as they tend to have more decision-making power and can therefore implement changes and ideas brought about through the scenario development process more effectively [42].

It has been argued that regardless of which actor is being focused on, there are three key points to bear in mind in terms of targeting actors with information. A clear plan of action needs to be laid out and followed up on. The information needs to be shared in a manner that is non-threatening, interactive and flexible. Scenarios can be disseminated by tapping into existing networks and events such as management meetings, seminars and the media [42].

In addition, the way that scenarios are packaged and communicated is important [42]. There is a whole host of ways that information can be packaged and disseminated. There can be face-to-face dissemination [23], where scenarios are verbally presented at workshops, conferences, public gatherings, business breakfasts, and corporate events. Style of presentation is crucial in this regard. The presentations need to be simple, clear and memorable. The presenter needs to be engaging and open to feedback from the audience [35]. Radio or television documentaries can also be utilised to disseminate scenario ideas and generate public debate [36, 40, 42].

Another option is to publish the scenarios in a written format. A range of media can be used. The scenarios can be published in books, illustrated pamphlets [23] and newspapers [48]. Cartoon artists can be brought on board to illustrate the scenarios. Magazines and web pages can also be targeted. Written documentation about scenarios has proved to be a successful model. For example, Sunter and Illbury's 'The Mind of the Fox: Scenario Planning in Action' [49] is popular reading material and widely distributed.

Finally, it is crucial that the scenario products are seen as legitimate from the start of the scenario development process. They need to have buy-in from influential people involved in the issue that the scenarios explore. This legitimacy is generated by ensuring that the facilitators of the process as well as the scenario team are respected. Although a range of actors must be included in the scenario development process, and must be targeted in the dissemination process, it remains important to include high level and well-connected people in the team as it is often these individuals who will provide the 'insider' links for scenarios to be heard and disseminated through channels of influence [42]. If these strategic individuals cannot be made part of the team, they need to be made aware of and kept informed about the scenario development process to secure their interest and support [35].

Dissemination is not without its challenges. It is challenging to disseminate in a way that suits and reaches a diverse audience with different languages, levels of education, varied professional backgrounds and cultures. Another challenge of ensuring the uptake of scenarios (and research in general) is that dissemination is often not part of the project planning process, and as a result funding often runs out before scenario uptake and use can be promoted [48]. Also, depending on how it is done, the dissemination of scenarios can be very expensive [35].

Over and above the specific processes linked to the impact of scenarios on the scenario team and broader society there are some general points that are important to bear in mind when planning for impact in relation to scenario products.

Firstly, when starting the scenario development process, it is important to be clear about the purpose of the process one is undertaking and designing it accordingly. What are the intended outcomes of the process? Who is the process meant to influence and what product(s) will be necessary for this to happen [35]? Essentially scenarios need to fill a strategic gap or opportunity in society [50].

Secondly, questions also need to be asked about the timing of the scenario development process. Is there likely to be sufficient recognition among the intended target audience(s) that the problem being addressed is important and that the process is therefore potentially beneficial? Is the political environment such that intended target audience(s) will be responsive to fresh, unorthodox thinking [35,42]?

Thirdly, attention also needs to be paid to the legitimacy of those financing and promoting the process, and the credibility of the project team developing the scenarios in the eyes of both the sponsors and the target audience(s) [35, 40, 42].

CONCLUSIONS

In conclusion, it seems that since the initial High Road/Low Road scenarios were developed, scenario development has taken root in South Africa, with several follow up scenarios having been developed since [28].

This development suggests that South African decision-makers must deem scenario development to be of considerable importance and utility, as it is often government or government-related institutions that develop or commission new sets of scenarios. These subsequent scenarios seem to mirror their predecessors with their snappy titles and straightforward structure and certainly have the potential to inspire decision-makers with regard to their planning activities [28].

Based on the discussion and reflections above, scenario development should involve a focus on dissemination and impact from the onset of the scenario process. Impact can happen at the level of participants in the scenario development process as they are exposed to new ideas and start adopting a new way of thinking about current issues of importance. These ideas have the potential to slowly infiltrate the networks of these participants and to also influence their thinking. At the same time, it is important to have a strong dissemination process in order to reach as many people as possible beyond the project team. The High Road/Low Road presentations are an example of a highly effective dissemination process made possible by an engaging speaker and interesting topic that was clearly and simply brought across to a wide range of audiences. Another key method of dissemination is to raise awareness about where the scenarios can be found and to make it easy for people to access them. The open access route followed by the Dinokeng scenario team is a good example of a scenario document that is easily available on a website, accompanied by much useful background information. It is this dissemination phase that has been lacking in the Water Sector Institutional Landscape by 2025 scenarios, and a follow up process is needed to plan how more people could be made aware of these scenarios and their usefulness to decision-making and planning in the South African water sector.

It is also important to keep in mind that scenarios are likely to have a higher impact if they are developed with the intention of identifying or solving particular problems [51]. If there is an intended target audience with particular information needs from the beginning of the scenario process, the scenario team will be able to keep this in mind when developing the scenarios. This will also ensure more effective uptake of the scenarios as pre-defined end-users exist. In the water sector, for instance, it could be effective for decision-makers who are grappling with a particular issue to solicit scenario inputs to aid them in making decisions regarding that issue.

In terms of future research, three areas come to mind based on what has been discussed in this chapter:

Firstly, a large scale study (mostly comprising of interviews) is needed to understand in greater detail the impact of scenarios on scenario participants, society and government planning processes [28]. Much of what has been argued to date in terms of the impact of scenarios has been on the basis of inference and assumptions. It would be interesting, though admittedly also very difficult, to try to substantiate views around the impact of scenarios with empirical evidence.

Secondly, it would be important to study how a scenario team would know that the timing is right to come up with and disseminate a new set of scenarios. It is reasonably easy to see that scenarios would have been important for particular moments in history, for example the political transition in South Africa, but it is considerably more difficult to determine when there may be an ideal window of opportunity in future in which scenarios may make an impact. It may also be important to determine which factors other than and in support of ideal timing would be important for scenarios to achieve impact.

Thirdly, building on this chapter, it would be important to determine how best to ensure that scenarios can become more useful and practical to policy-makers and other end-users. How can scenario teams ensure that end-users know how best they may use scenarios in order to influence their future planning? The issue of providing navigation to and between different scenarios and future outcomes is important in this regard.

Clearly, scenario development is a useful process to help decision-makers cope with and plan amidst uncertainty. Particularly in the context of the South African water sector, it is important to recognise that uncertainty is deepening in many ways given the impending presence of multiple stressors such as climate change, basin closure, growing populations, migration movements and a growing economy. These stressors, along with the institutional fluctuations and changes within the water sector itself, make it increasingly important for decision-makers to work with scenarios to help them to plan sensibly and creatively despite uncertainty. However, in order for scenarios to be useful it is important to plan for and carefully think about how to maximise their impact.

ACKNOWLEDGEMENTS

The authors would like to acknowledge the CSIR's librarian, Engela van Heerden, for her excellent work in sourcing a large amount of relevant literature that contributed to this chapter. They would also like to acknowledge Wilma Strydom for her valuable review comments on this chapter. Finally, the authors would like to acknowledge the Water Research Commission (WRC), the organisation that funded the development of the Water Sector Institutional Landscape by 2025 scenarios. The learning that the project team gained from this project was a key input into this chapter.

REFERENCES

1. Bogardi JJ, Dudgeon D, Lawford R, Flinkerbusch E, Meyn A, Pahl-Wostl C, Vielhauuer K and Vörösmarty C. Water Security for a Planet Under Pressure: Interconnected Challenges of a Changing World Call for Sustainable Solutions. Current Opinion in Environmental Sustainability 2012; 4 35–43.
2. Slaughter RA. Welcome to the Anthropocene. Futures 2012; 44 119–126.
3. WWAP (World Water Assessment Programme). The United Nations World Water Development Report 4: Managing Water under Uncertainty and Risk. Paris: UNESCO; 2012.
4. UN (United Nations). Declaration of the United Nations Conference on the Human Environment. Stockholm, 5-16 June 1972.
5. Brundtland GH. Address by Mrs Gro Harlem Brundtland, Chairman at the Closing Ceremony of the Eighth and Final Meeting of the World Commission on Environment and Development 27 February 1987. Tokyo, Japan.
6. WMO (World Meteorological Organisation). International Conference on Water and the Environment: Development Issues for the 21st Century. Geneva: ICWE Secretariat; 1992.
7. UN (United Nations). Agenda 21. United Nations Conference on Environment and Development Rio de Janeiro, Brazil, 3 to 14 June 1992.
8. Spangenberg JH, Pfahl S, Deller K. Towards Indicators for Institutional Sustainability: Lessons From an Analysis of Agenda 21. Ecological Indicators 2002; 2 61–77.
9. UN (United Nations). World Development Report, 1980. Washington, D.C.: The World Bank; 1980.
10. UN (United Nations). World Development Report, 1985. New York: Oxford University Press; 1985.

REFERENCES

11. UN (United Nations). World Development Report, 1990: Poverty. New York: Oxford University Press; 1985.
12. UN (United Nations). World Development Report, 1995: Workers in an Integrating World. New York: Oxford University Press; 1995.
13. World Bank. Indicators. http://data.worldbank.org/indicator. (accessed 30 July 2012).
14. Bates C, Kundzewicz ZW, Wu S and Palutikof JP., editors. Climate Change and Water. Technical Paper of the Intergovernmental Panel on Climate Change. Geneva: IPCC Secretariat; 2008.
15. Davis CL. Climate Risk and Vulnerability: A Handbook for Southern Africa. Pretoria: Council for Scientific and Industrial Research; 2011.
16. Tewari DD. A Detailed Analysis of the Evolution of Water Rights in South Africa: An Account of Three and a Half Centuries from 1652 AD to the Present. Water SA 2009; 35(5) 693-710.
17. Gleick PH. The World's Water, 1998-1999. Washington DC: Island Press; 1998.
18. DWAF (Department of Water Affairs and Forestry). National Water Act of South Africa. Act No. 36 of 1998. Pretoria: Department of Water Affairs and Forestry; 1998.
19. DWAF (Department of Water Affairs and Forestry). National Water Resources Strategy – First Edition. Pretoria: Department of Water Affairs and Forestry; 2004.
20. DWA (Department of Water Affairs). Annual Report of the Department of Water Affairs Vote 37: 1 April 2010 To 31 March 2011. http://www.pmg.org.za/minutes/30. (accessed 6 August 2012).
21. Emerson JW, Hsu A, Levy MA, de Sherbinin A, Mara V, Esty DC, Jaiteh M. Environmental Performance Index and Pilot Trend Environmental Performance Index. New Haven: Yale Center for Environmental Law and Policy; 2012.
22. Green GC, Day JA, Mitchell SA, Palmer C, Laker MC, Buckley CA. Water Research Commission 40-Year Celebration Conference - Syntheses of Themed Sessions. Water SA 2011; 37(5) 609-618.
23. Claassen M, Funke N, Nienaber S. 2011. The Water Sector Institutional Landscape by 2025 Technical Report. WRC Report No. 1841/1/11. Pretoria: Water Research Commission; 2011.
24. Mietzner D, Reger G. EU-US Seminar: New Technology Foresight, Forecasting and Assessment Methods. Seville; 13-14 May 2004.
25. Landeta J. Current Validity of the Delphi Method in Social Sciences. Technological Forecasting and Social Change 2006; 73 467-482.

26. Daum J. How Scenario Planning Can Significantly Reduce Strategic Risks and Boost Value in the Innovation Chain. The New Economy Analyst Report 8 September 2001.
27. Sunter C. The World and South Africa in the 1990s. Cape Town: Human Rousseau Tafelberg; 1987.
28. Galer G. Scenarios of Change in South Africa. The Round Table 2004; 93(375) 369-383.
29. DACST (Department of Arts, Culture, Science and Technology). South African National Research and Technology Foresight Project. Pretoria: DACST; 1999.
30. Dinokeng team. 3 Futures for South Africa. www.dinokengscenarios.co.za (accessed 10 August 2012).
31. (WBCSD) World Business Council for Sustainable Development. Business in the World of Water: WBCSD Water Scenarios to 2025. www.wbcsd.org/web/H2OScenarios.htm (accessed 30 July 2012).
32. GRA (Global Research Alliance) Science and Technology-Based Water Scenarios for Sub-Saharan Africa. www.research-alliance.net. (accessed 6 August 2012).
33. Peterson GD, Cumming GS, Carpenter SR. Scenario Planning: A Tool for Conservation in an Uncertain World. Conservation Biology 2003; 17(2) 358-366.
34. Finlev T. Future Peace: Breaking Cycles of Violence through Futures Thinking. Journal of Futures Studies 2012; 16(3) 47-62.
35. Segal N. Breaking the Mould: The Role of Scenarios in Shaping South Africa's Future. Stellenbosch: SUN Press; 2007.
36. Kahane A. The Mont Fleur Scenarios: What Will South Africe Be Like in the Year 2002? Deeper News 1992, 7(1) 1-22.
37. Maddison S, Cronin D, Williams S, Coggan R. Democratic Dialogue: Finding the Right Model for Australia. Indigenous Policy and Research Unit, Discussion Paper No.1. Sydney: University of New South Wales; 2009.
38. Claassen M, Funke N, Nienaber S. The Water Sector Institutional Landscape by 2025. WRC Report No. TT 514/11. Pretoria: Water Research Commission; 2011.
39. Karuri-Sebina G, Rosenzweig L. A Case Study on Localising Foresight in South Africa: Using Foresight in the Context of Local Government Participatory Planning. Foresight 2007, 14(1) 26-40.
40. Strydom WF, Funke N, Nienaber S, Nortje K, Steyn M. Evidence-based Policy-making: A Review. South African Journal of Science 2010; 106(5/6) 249 – 334.

41. Funke N, Nienaber S, Henwood R. Scientists as Lobbyists? How Science Can Make its Voice Heard in the South Africa Policy-Making Arena. International Journal of Public Affairs 2011; 10(102) 421 – 443.
42. Funke N, Nienaber S. Promoting Uptake and Use of Conservation Science in South Africa by Government. Water SA 2012; 38(1) 105-114.
43. Lebel L, Thongbai P, Kok K, Agard JBR, Bennett EM, Biggs R, Ferreira M, Filer C, Gokhale Y, Mala W, Rumsey C, Velarde SJ, Zurek M, Blanco H, Lynam T, Tianxiang Y. Sub-global Scenarios. In: Capistrano D, Lee M, Raudsepp-Hearne C, Samper C. (eds.) Ecosystems and Human Wellbeing: Multi-scale Assessments. Findings of the Subglobal Assessments Working Group of the Millennium Ecosystem Assessment. Washington D.C.: Island Press; 2005. p229-259.
44. Ringland G. Scenario Planning, 2nd edition. England: John Wiley & Sons; 2006.
45. Audouin M, Preiser R, Nienaber S, Downsborough L, Lanz J, Mavengahama S. Exploring the Logic of Complexity for Researching Social-Ecological Systems. Ecology and Society In Press.
46. Ogilvy JA. Creating Better Futures: Scenario Planning as a Tool for a Better Tomorrow. New York: Oxford University Press; 2002.
47. Jacobs I, Nienaber S. Water Without Borders: Transboundary Water Governance and the Role of the 'Transdisciplinary Individual' in Southern Africa. Water SA (WRC 40 Year Celebration Special Edition 2011; 37(5) 665 – 678.
48. Communication with Clem Sunter. 26 July 2012. Pretoria.
49. Illbury C, Sunter C. The Mind of a Fox. Cape Town: Human & Rousseau/Tafelberg; 2001.
50. Johnson G, Scholes K, Whittington R. Exploring Corporate Strategy: Text and Cases. New Jersey: Prentice Hall, Financial Times Press; 2008.
51. Bohenksy E, Reyers B, Van Jaarsveld AS. Future Ecosystem Services in a Southern African River Basin: a Scenario Planning Approach to Uncertainty. Conservation Biology 2006; 20(4) 1051-1061.

CITATION

Nikki Funke, Marius Claassen and Shanna Nienaber (2013). Development and Uptake of Scenarios to Support Water Resources Planning, Development and Management – Examples from South Africa, Water Resources Planning, Development and Management, Prof. Ralph Wurbs (Ed.), ISBN: 978-953-51-1092-7, InTech, DOI: 10.5772/52577. Available from: http://www.intechopen.com/books/water-resources-planning-development-and-management/development-and-uptake-of-scenarios-to-support-water-resources-planning-development-and-management-e

Index

A
Agro-hydrological modeling system (ACRU), 187
Analytic hierarchy process (AHP), 158
Artificial network, 60

B
Benchmark system, 3
Bifurcation, 118
Billion Cubic Meter (BCM), 185

C
Charge Coupled Device (CCD), 117, 121
Churchman–Ackoff method, 164
countries, 208, 210
Cubic meters per day (CMD, 170
Cyprus, 29, 30, 31, 32, 33, 34, 53, 54, 56

D
Dendritic network, 63
development, 207, 208, 209, 210, 211, 212, 213, 215, 219, 220, 221, 222, 223, 224, 225, 226, 227, 228, 230, 231, 232, 233, 234, 238
Diverse scientific field, 29
Drainage system, 117, 118, 119, 120, 132, 137, 138

E
Ecocene limestone, 189
economic, 207, 208, 209, 210, 213, 215, 217, 220, 221, 224
Electromagnetic, 29, 38, 41
environmental, 207, 208, 210, 215, 224
Evolutionary Algorithm, 64, 72, 87

Evolutionary algorithms (EAs), 60
Evolutionary Algorithms (EAs), 59, 62
Evolutionary Computation (EC), 62

F

Fayoum Governorate, 185, 186, 187, 190, 191, 193, 197, 198, 203, 205
Fayoum Governorate face, 185

G

Ground penetrating radar (GPR), 94, 96

H

Heat-transfer, 139, 140, 142, 144, 146, 148, 149
Horizontal expansion project, 186
Human intervention, 60
Hydraulic, 153, 154, 155, 156, 157, 161, 162, 163, 179
Hydraulic device, 61, 68
Hydraulic performance, 3, 4, 12, 19
Hydroinformatic, 64, 67
Hydrophilic materials, 141
Hyperheuristics, 59, 60

I

industrial, 207, 209, 223
Infiltration basin, 93, 94, 114
Infrastructural asset management (IAM), 157

Integrated bioinspired, 139, 151
international, 207, 213

L

Linear extension of the Yule process (LEYP), 157
Lithological heterogeneity, 93, 95, 101
Low impact development, 1, 2

N

National Research Foundation (NRF), 180
National Water Resources Plan in Egypt (NWRP), 188
Nodal serviceability (N_S), 162, 163, 174

O

Optimisation process, 65, 72
Optimisation scale, 65

P

Pipes, 139, 140, 141, 142, 143, 144, 146, 147, 148, 149, 150, 151
Principal Component Analysis (PCA), 37
Probabilistic neural network (PNN), 155
prosperity, 207, 214

Q

Quarun lake, 188
Quarun Lake, 185, 186, 187, 189, 190, 201, 203, 204

Index

R
Radiometric calibration, 37
Remote sensing, 30, 31, 33, 37, 46, 48, 49, 54, 55, 56, 57

S
Sedimentological, 93, 94, 95, 98, 99, 100, 101, 107, 112, 113
Semi-synthetic, 3
significantly, 208, 209
Soil heterogeneity, 93, 95, 109, 111, 112, 113, 114
Spectroradiometric, 31, 37, 38, 39, 54
Strong-base-oxidized, 139
Swirl generation vanes (SGV), 117

U
Unintended isolation (UI), 172
Urban water management, 1, 2, 23, 24

Urban water system, 1, 2, 3, 4, 6, 22

V
variable, 208, 209, 210, 227
Vegetation indices (VI), 30
Virtual Infrastructure Benchmarking, 2, 3, 22

W
Water distribution network (WDN), 62
Water distribution networks (WDN), 59
Water evolution and planning (WEAP), 187
Water infiltration, 93, 94, 95, 99, 107, 108, 109, 111, 113
Weighted utopian approach (WUA), 167
Work packages (WP),, 156